Distributed Adaptive Consensus Control of Uncertain Multi-agent Systems

Multi-agent systems are special networked systems full of research interest and practical sense, which are abundant in real life, ranging from mobile robot networks, intelligent transportation management, to multiple spacecraft, surveillance and monitoring. Consensus control is one of the most typical and hot research issues for multi-agent systems. **Distributed Adaptive Consensus Control of Uncertain Multi-agent Systems** provides innovative technologies to design and analyze distributed adaptive consensus for multi-agent systems with model uncertainties.

Based on the basic graph theory and adaptive backstepping control, this monograph:

- Describes the state of the art on distributed adaptive control, finite-time consensus control and event-triggered consensus control
- Studies distributed adaptive consensus under *directed* communication graph condition: the methods with linearly parametric reference, hierarchical decomposition, and design of auxiliary filers
- Explores adaptive *finite-time* consensus for uncertain nonlinear systems
- Considers distributed adaptive consensus with *event-triggered* communication via state feedback and output feedback
- Investigates distributed adaptive formation control of *nonholonomic mobile robots* with *experimental verification*
- Provides distributed adaptive attitude synchronization control schemes for multiple *spacecraft* with event-triggered communication

Distributed Adaptive Consensus Control of Uncertain Multi-agent Systems can help engineering students and professionals to efficiently learn distributed adaptive control design tool for handling uncertain multi-agent systems with directed communication graph, guaranteeing finite-time convergence and saving communication resources.

Adaptive and Fault-Tolerant Control of Underactuated Nonlinear Systems
Jiangshuai Huang, Yong-Duan Song

Discrete-Time Recurrent Neural Control
Analysis and Application
Edgar N. Sánchez

Control of Nonlinear Systems via PI, PD and PID
Stability and Performance
Yong-Duan Song

Multi-Agent Systems
 Platoon Control and Non-Fragile Quantized Consensus
Xiang-Gui Guo, Jian-Liang Wang, Fang Liao, Rodney Swee Huat Teo

Classical Feedback Control with Nonlinear Multi-Loop Systems
With MATLAB® and Simulink®, Third Edition
Boris J. Lurie, Paul Enright

Motion Control of Functionally Related Systems
Tarik Uzunović and Asif Sabanović

Intelligent Fault Diagnosis and Accommodation Control
Sunan Huang, Kok Kiong Tan, Poi Voon Er, Tong Heng Lee

Nonlinear Pinning Control of Complex Dynamical Networks
Edgar N. Sanchez, Carlos J. Vega, Oscar J. Suarez and Guanrong Chen

Adaptive Control of Dynamic Systems with Uncertainty and Quantization
Jing Zhou, Lantao Xing and Changyun Wen

Robust Formation Control for Multiple Unmanned Aerial Vehicles
Hao Liu, Deyuan Liu, Yan Wan, Frank L. Lewis, Kimon P. Valavanis

Variable Gain Control and Its Applications in Energy Conversion
Chenghui Zhang, Le Chang, Cheng Fu

Manoeuvrable Formation Control in Constrained Space
Dongyu Li, Xiaomei Liu, Qinglei Hu, and Shuzhi Sam Ge

Distributed Adaptive Consensus Control of Uncertain Multi-agent Systems
Wei Wang, Jiang Long, Jiangshuai Huang, Changyun Wen

For more information about this series, please visit: https://www.crcpress.com/
Automation-and-Control-Engineering/book-series/CRCAUTCONENG

Distributed Adaptive Consensus Control of Uncertain Multi-agent Systems

Wei Wang, Jiang Long, Jiangshuai Huang,
and Changyun Wen

CRC Press
Taylor & Francis Group
Boca Raton London New York

CRC Press is an imprint of the
Taylor & Francis Group, an **informa** business

First edition published 2025

by CRC Press
2385 NW Executive Center Drive, Suite 320, Boca Raton FL 33431

and by CRC Press
4 Park Square, Milton Park, Abingdon, Oxon, OX14 4RN

CRC Press is an imprint of Taylor & Francis Group, LLC

© 2025 [Wei Wang, Jiang Long, Jiangshuai Huang, Changyun Wen]

ISBN: 978-1-032-49546-0 (hbk)
ISBN: 978-1-032-49549-1 (pbk)
ISBN: 978-1-003-39437-2 (ebk)

DOI: 10.1201/9781003394372

Typeset in Nimbus Roman
by KnowledgeWorks Global Ltd.

Publisher's note: This book has been prepared from camera-ready copy provided by the authors.

Contents

SECTION I   Consensus Control under Directed Graph

                Control 55

                4.1   Problem Formulation 55
                4.2   Design of Distributed Filters 56
                4.3   Design of Local Adaptive Controllers 58
                4.4   Stability and Consensus Analysis 60
                4.5   Simulation Results 63
                4.6   Notes 65

Chapter 5   Hierarchical Decomposition-Based Distributed Adaptive
                Consensus Control 66

                5.1   Problem Formulation 66
                5.2   Hierarchical Design of Distributed Adaptive Controllers 67
                       5.2.1   Hierarchical Decomposition 67
                       5.2.2   Design of Distributed Adaptive Controllers 69
                5.3   Stability and Consensus Analysis 73
                5.4   Simulation Results 74
                5.5   Notes 77

SECTION II   Finite-Time Consensus Control

                Mechanical Systems 81

                6.1   Preliminaries 82
                6.2   Problem Formulation 83
                6.3   Adaptive Finite-Time Consensus Controller Design 83
                6.4   Stability Analysis 88
                6.5   Transient Performance Analysis 92
                6.6   Simulation Results 95
                6.7   Notes 99

SECTION III   Consensus Control with
Event-Triggered Communication

                Case 103

                7.1   Problem Formulation 104
                7.2   Preliminary Control Design with Continuous
                       Communication 104
                7.3   Control Design with Event-Triggered Communication 107
                       7.3.1   Design of Triggering Condition 108

SECTION IV   Applications

Preface

Multi-agent systems are special networked systems full of research interest and practical sense, which are abundant in real life, ranging from mobile robot networks and intelligent transportation management, to multiple spacecraft, surveillance, and monitoring. Consensus control is one of the most typical and hot research issues for multi-agent systems. It aims to achieve an agreement on the states/outputs for all the agents, by designing a local controller for each agent. Two control frameworks are commonly adopted, i.e., centralized framework and distributed framework. The former is developed based on the assumption that a central station is available and sufficiently powerful to control the entire group of agents. By contrast, the distributed framework can remove the requirement of central station by effectively utilizing information transmission among different agents, while at the cost of becoming far more complex in structure and organization. Although both frameworks are practical depending on different situations and conditions of real applications, the distributed framework is often more promising in scenarios with various inevitable physical constraints such as limited resources and energy, short wireless communication ranges, and large number of agents to control.

Currently, a great number of distributed consensus control methodologies have been proposed for precisely known linear and first-order or second-order nonlinear multi-agent systems. However, the results on more general high-order nonlinear multi-agent systems are still limited. Besides, it is well known that uncertainties including unknown system parameters and external disturbances inevitably exist in practice. Therefore, developing distributed consensus control strategies for uncertain high-order nonlinear multi-agent systems is of great significance. Adaptive control is an effective approach to handle model uncertainties as it can offer online estimation for uncertain parameters. However, valid distributed consensus control schemes employing the adaptive techniques are still limited, especially for the cases with directed communication graphs. The main reason is that directed graph is associated with the asymmetric Laplacian matrix, which will bring about new challenges to design distributed adaptive laws based on Lyapunov stability theory.

An important performance indicator for consensus problem is the convergence rate. Most of the current consensus control schemes for multi-agent systems achieve asymptotic convergence. In other words, the convergence rate at best is exponential. This implies that it takes an infinite amount of time for the tracking errors to converge to the origin. In practice, it is more significant if consensus can be reached within finite time. In addition, finite-time control schemes possess better disturbance rejection properties. Except for improving control performance, how to relieve the communication burden among connected agents with limited communication resources is another important issue. To this end, event-based communication mechanism is naturally introduced in designing consensus control schemes, such that the communication among agents is activated only when some predefined triggering condition is triggered.

This monograph details a series of innovative technologies of designing and analyzing high-order multi-agent systems with uncertainties, finite-time convergence performance, event-triggered communication under directed graph condition. State feedback and output feedback consensus control schemes are presented. Moreover, the proposed control schemes are utilized in two common applications in nonholonomic mobile robots and multiple spacecraft networks. These results are given in four parts.

In the first part of this monograph, the distributed adaptive consensus tracking control problem for high-order nonlinear multi-agent systems with unmatched system parameters and directed graph will be investigated in Chapter 3, under the assumption that the desired trajectory takes the linearly parametric form. In order to relax such assumption, auxiliary filter-based and hierarchical decomposition-based adaptive consensus tracking control schemes will be presented in Chapters 4 and 5, respectively.

In the second part of this monograph, a distributed adaptive finite-time consensus control scheme for a class of uncertain nonlinear mechanical systems will be presented in Chapter 6 to improve the consensus convergence rate. With the presented control scheme, the outputs of all the agents can reach leaderless consensus within finite time.

In the third part of this monograph, some advances in distributed adaptive consensus control under event-triggered communication condition are presented to save communication resources among the connected agents. First, the leader-following and leaderless consensus control problems for a class of uncertain high-order nonlinear multi-agent systems will be, respectively, discussed in Chapters 7 and 8 based on state feedback control. Then, the consensus tracking control problem for uncertain linear multi-agent systems with unknown system parameters and event-triggered communication will be investigated in Chapter 9 via output feedback control.

In the fourth part of this monograph, the consensus control techniques will be utilized to solve the formation control problem for nonholonomic mobile robots in Chapter 10 and attitude synchronization problem for multiple spacecraft in Chapter 11.

This monograph is helpful to learn and understand the design and analysis of distributed adaptive coordinated control systems to achieve consensus of uncertain nonlinear multi-agent systems. It can be used as a reference book or a textbook on distributed adaptive control theory and applications for students with the background in control systems. The monograph is also intended to introduce researchers and practitioners to the area of multi-agent systems involving the treatment on model uncertainties, directed graph, finite-time convergence rate, and event-triggered communication. Researchers, graduate students, and engineers in the fields of electrical engineering, control, applied mathematics, computer science, and others will benefit from this monograph.

We are grateful to Beihang University (China), Northwestern Polytechnical University (China), Chongqing University (China), and Nanyang Technological University (Singapore) for providing plenty of resources for our research work. Wei Wang appreciates and acknowledges National Natural Science Foundation of China for their support with Grants 62373019, 62022008 and 61973017. Jiang Long appreciates and acknowledges National Natural Science Foundation of China for their support with Grant 62203361. We express our deep sense of gratitude to our beloved families who have made us capable enough to write this book. Wei Wang is very grateful to her parents, Xiaolie Wang and Minna Suo, her husband, Qiang Wu, and her daughter, Huanxin Wu, for their care, understanding, and constant encouragement. Jiang Long is greatly indebted to his parents, Siquan Long and Yonghui He, his wife, Juan Jin, and his sister Xin Long for their constant support throughout these years. Jiangshuai Huang is very grateful to his parents, Zifu Huang and Xueyuan Zhang, his wife, Xiangjun Wang and their precious son, Xingjian Huang, for their unwavering support and cherished companionship. Changyun Wen is greatly indebted to his wife, Xiu Zhou and his children Wen Wen, Wendy Wen, Qingyun Wen and Qinghao Wen for their constant invaluable support and assistance throughout these years.

Finally, we thank the entire team of CRC Press for their cooperation and great efforts in transforming the raw manuscript into a book.

Wei Wang
Beihang University, China

Jiang Long
Northwestern Polytechnical University, China

Jiangshuai Huang
Chongqing University, China

Changyun Wen
Nanyang Technological University, Singapore

Authors

Wei Wang received her B.Eng degree in Electrical Engineering and Automation from Beihang University (China) in 2005, M.Sc degree in Radio Frequency Communication Systems with Distinction from University of Southampton (UK) in 2006 and Ph.D degree from Nanyang Technological University (Singapore) in 2011. From January 2012 to June 2015, she was a Lecturer with the Department of Automation at Tsinghua University, China. Since July 2015, she has been with the School of Automation Science and Electrical Engineering, Beihang University, China, where she is currently a Full Professor. Her research interests include adaptive control of uncertain systems, distributed cooperative control of multi-agent systems, secure control of cyber-physical systems. Prof. Wang received Zhang Si-Ying Outstanding Youth Paper Award in the 25th Chinese Control and Decision Conference (2013) and the First Prize of Science and Technology Progress Award by Chinese Institute of Command and Control (CICC) in 2018. She is the Principle Investigator for a number of research projects including the Distinguished Young Scholars of the National Natural Science Foundation of China (2021-2023). She has been serving as Associate Editors for the IEEE Transactions on Industrial Electronics, ISA Transactions, IEEE Open Journal of Circuits and Systems, Journal of Control and Decision, Journal of Command and Control.

Jiang Long received the B.Eng. degree in electrical engineering and automation from the Civil Aviation University of China, China, in July 2015 and the Ph.D. degree in control theory and control engineering from Beihang University, China, in November 2021. Since 2022, he is an Associated Professor with the School of Computer Science, Northwestern Polytechnical University, Xi'an, China. His research interests include adaptive control, nonlinear systems control, distributed adaptive control, event-triggered control. He has been serving as a Reviewer for Automatica, IEEE Transactions on Automatic Control, IEEE Transactions on Cybernetics, IEEE Transactions on Industrial Electronics, Systems & Control Letters, etc.

Jiangshuai Huang received the B.Eng. and M.Sc. degrees in automation from Huazhong University of Science and Technology, Wuhan, China, in 2007 and 2009, respectively, and the Ph.D. degree from the School of Electrical and Electronic Engineering, Nanyang Technological University, Singapore, in 2015. He was a Research Fellow with the Department of Electricity and Computer Engineering, National University of Singapore from August 2014 to September 2016. He has been with the School of Automation, Chongqing University since October 2016, where he is currently a Full Professor. His research interests include adaptive control, nonlinear systems control, underactuated mechanical system control, and multiagent system control. Prof. Huang received Zhang Si-Ying Outstanding Youth Paper Award in the 25th Chinese Control and Decision Conference, the First Prize of Science and Technology Progress Award by Chinese Institute of Command and Control (CICC) in 2018, the First Prize of Natural Science Award by Government of Chongqing in 2020 and the First Prize of Natural Science Award by Chinese Association of Automation in 2021. He is an Associate Editor for the IEEE Transactions on Cybernetics and IEEE Transactions on Industrial Electronics.

Changyun Wen received his B.Eng from Xi'an Jiaotong University, China in 1983 and Ph.D from the University of Newcastle, Australia in 1990. From August 1989 to August 1991, he was a Postdoctoral Fellow at the University of Adelaide, Australia. He is presently a Professor at Nanyang Technological University, Singapore. He is a Fellow of the Academy of Engineering, Singapore, a Fellow of IEEE, was a Member of the IEEE Fellow Committee from Jan 2011 to Dec 2013 and a Distinguished Lecturer of IEEE Control Systems Society from 2010 to 2013. Currently, he is a co-Editor-in-Chief of IEEE Transactions on Industrial Electronics, an Associate Editor of Automatica and the Executive Editor-in-Chief of Journal of Control and Decision. Staring from January 2000, he also served as an Associate Editor of IEEE Transactions on Automatic Control, IEEE Transactions on Industrial Electronics and IEEE Control Systems Magazine. He has been actively involved in organizing international conferences playing the roles of General Chair, TPC Chair, General Advisor ect. He was the recipient of several outstanding awards, including the IES Prestigious Engineering Achievement Award from the Institution of Engineers, Singapore in 2005, and Best Paper Award of IEEE Transactions on Industrial Electronics in 2017.

1 Introduction

In recent years, multi-agent systems have gained significant attention in control theory and control engineering due to the promising applications such as in mobile robot networks, intelligent transportation management, and multiple spacecraft. Consensus control is one of the most typical and hot research issues for multi-agent systems. It aims to achieve an agreement on the states/outputs for all the agents by designing a local controller for each agent. A great number of distributed consensus control methodologies have been proposed for precisely known linear and first-order or second-order nonlinear multi-agent systems. However, the results on more general high-order nonlinear multi-agent systems with model uncertainties and directed graph are still limited. Besides, how to improve the consensus convergence rate and save communication resources of multi-agent systems have not been extensively explored.

1.1 ADAPTIVE CONTROL

Adaptive control is a design technique that involves dynamically self-tunning the control parameters based on performance error related information in order to optimize system behavior within its environment. This approach enables the achievement of various objectives, including system stability, desired output tracking with guaranteed steady state accuracy, and improved transient performance. Since its inception in the early 1950s, adaptive control has emerged as a research area of significant theoretical and practical importance. Notably, the development of autopilot systems for high-performance aircraft has been a key driving force behind extensive research in adaptive control [47]. Over the course of nearly seven decades, numerous adaptive control design techniques have been proposed to address different classes of systems and tackle different challenges. Among the commonly employed conventional adaptive control methods are Model Reference Adaptive Control (MRAC) [82, 93, 141], system and parameter identification-based schemes [3, 94], and adaptive pole placement control [31, 32]. In the 1980s, several modification techniques were introduced to enhance the robustness of adaptive controllers against unmodeled dynamics, disturbances, and modeling errors. These techniques include normalization [92, 96], dead-zone [34, 53], switching σ-modification [46], and parameter projection [85, 140, 152]. In the early 1990s, adaptive backstepping control [56] emerged as a solution for controlling certain classes of nonlinear plants with unknown parameters. The introduction of tuning functions contributed to enhanced transient performance of adaptive control systems. The breakthroughs mentioned above represent only a fraction of the notable advancements in the field of adaptive control. For more comprehensive literature reviews on conventional adaptive control, interested readers can refer to works such as [4,35,47,86,125], as well as other related textbooks and survey papers.

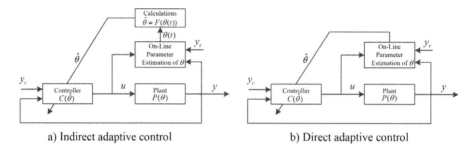

a) Indirect adaptive control   b) Direct adaptive control

FIGURE 1.1   The block diagram of direct and indirect adaptive control.

One notable aspect that has contributed to the rapid advancement of adaptive control is its ability to effectively handle systems with unknown parameters. Unlike many non-adaptive techniques, adaptive controllers are not typically designed based on prior knowledge of the uncertainties present in the systems. Instead, an adaptive controller is normally designed by combining parameter update law and control law to achieve desired performance. The parameter update law, also known as the parameter estimator, plays a crucial role in the adaptive control framework. It provides an adaptation mechanism for adjusting the parameters of the controller in real time at each time instant [125]. By continuously updating these adjustable parameters based on the available information, the adaptive controller can adapt to changing conditions and mitigate the effects of unknown parameters, thus enhancing its overall performance.

In the past, adaptive control techniques were commonly categorized as either direct or indirect based on the approach used to obtain the controller parameters. The block diagram of direct and indirect adaptive control is provided in Figure 1.1 [48]. In indirect adaptive control, the controller parameters are computed based on the estimated system parameters. On the other hand, direct adaptive control involves estimating the controller parameters directly without intermediate calculations. Both indirect and direct adaptive control techniques share a common principle known as the certainty equivalence principle. This principle means that the controller structure is designed as if all estimated parameters were true, to achieve desired performances.

1.2   CONSENSUS CONTROL

In the early stages, humans were inspired by various biological phenomena in nature, such as ants collaborating to transport food, fish swimming in coordinated patterns, and geese migrating in organized formations. These observations led to the recognition that collective intelligent behavior, which surpasses the capabilities of individual entities, can emerge through the cooperation of multiple individuals. Over the past two decades, collective intelligent behavior has garnered significant attention from experts and scholars in physics, mathematics, biology, engineering, and social sciences. To facilitate research in this field, scholars have coined the term "intelligent agents" to describe individual entities that possess autonomous abilities to independently receive, transmit, and process information. They have also introduced network nodes to represent individual intelligent agents and topological structures to depict the interaction relationships among different agents. Furthermore, a collective system composed of multiple intelligent agents is defined as "Multi-Agent Systems" [11].

Consensus control, a crucial research issue in multi-agent systems, has attracted significant attention from experts and scholars due to its profound theoretical significance and practical value [11, 37, 100]. It aims to design cooperative control algorithms for multi-agent systems to achieve synchronization of outputs or states among all agents [111], which forms the foundation for studying complex cooperative control problems such as flocking, rendezvous, containment, and formation [111, 122]. The issues of flocking and rendezvous are specific examples of the consensus problem. The containment control problem addresses the consensus tracking problem with multiple leader agents [78,79]. In formation control, achieving consensus is a prerequisite for realizing formation configuration in multi-agent systems [7, 70]. Currently, research achievements in the field of consensus control have been successfully applied in various scenarios such as wireless sensor network localization [15], satellite attitude synchronization [8,76], unmanned aerial vehicles, mobile robots, and underwater vehicle formations [110, 138], as well as intelligent power distribution in smart grids [33].

There are primarily two types of consensus control schemes, i.e., centralized and distributed control schemes [11]. Centralized control schemes require global information about the quantity and graph of the multi-agent systems and utilize the states of all the agents. Essentially, centralized control schemes can be regarded as a direct extension of traditional single-system control approaches. Distributed control algorithms differ from centralized control algorithms as they do not rely on global information about the quantity and graph of the multi-agent systems. Instead, they only utilize the local and neighboring states of the agents. Thus, changes in the number of agents do not necessitate a redesign of the distributed control algorithm. Additionally, distributed control also offers advantages such as enhanced robustness, parallel processing capabilities, and increased reliability, which are particularly important for controlling ultra-large-scale intelligent agent systems. However, due to the inability to utilize global information of the multi-agent system, the design of distributed control algorithms often becomes structurally complex.

In the past two decades, scholars have conducted extensive research on multi-agent systems with precise models and have achieved fruitful results. For first-order integrator multi-agent systems, a theoretical framework for consensus control is established in [91] based on matrix theory and algebraic graph theory. It shows that the sufficient condition to achieve average consensus is that the underlying graph is strongly connected and balanced. In [110], it is concluded that the necessary and sufficient condition to achieve consensus with a fixed graph is that the underlying graph contains a directed spanning tree. The consensus state reached by the agents is determined by the initial states of those located at the root node. In [83, 109], a sufficient condition to achieve consensus in the presence of time-varying graphs is derived so that the joint connectivity graph has a directed spanning tree. Lately, the results are extended to second-order integrator multi-agent systems [104, 108, 156], higher-order integrator multi-agent systems [39, 49, 80, 113, 114, 149, 165], and general linear multi-agent systems [63, 65, 116, 126, 127] from various perspectives, such as dynamic feedback, switching graphs, communication noise, and time delays.

For nonlinear multi-agent systems, the consensus control problem is also extensively investigated. In [29] and [26], consensus control schemes for high-order nonlinear multi-agent systems with precisely known system dynamics are proposed in the presence and absence of a leader agent. In particular, for the case with a leader agent, if only a subset of agents can access the trajectory of the leader agent, a distributed consensus tracking control algorithm is designed based on backstepping control method [56]. For the case without a leader agent, a consensus control algorithm is developed to ensure that the outputs of all the agents can converge to a common value. In [107], the consensus control problem of Euler-Lagrange systems is investigated, where an anti-windup consensus control scheme is proposed to address input saturation. In [13], the consensus control problem of multiple oscillators is considered. A lower bound of the coupling gain that guarantees the existence of phase-locked states is derived if the natural frequencies of the oscillators can be arbitrarily chosen from a reasonable set of values. The result also provides sufficient conditions for individual oscillators to exponentially synchronize to the mean of the oscillators. In [20], the distributed consensus control problem for nonholonomic multi-robot systems with fixed graph is addressed, by using nonsmooth Lyapunov stability theory and algebraic graph theory. Lately, the results are extended to consider the case of dynamic graphs in [22]. In addition to the aforementioned references, the papers [14, 21, 27, 28, 70, 115] also investigate consensus control problems of nonholonomic mobile robots, Euler-Lagrange systems, or other complex dynamic multi-agent systems.

In summary, the research focus of consensus control problem has evolved from simple first-order integrator multi-agent systems to nonlinear multi-agent systems. For linear multi-agent systems, researchers generally employ matrix theory and algebraic graph theory to design consensus algorithms and analyze the stability of closed-loop systems. However, for nonlinear multi-agent systems with complex dynamics, modern stability analysis theories such as Lyapunov stability theory and passivity theory [100] are applied. With the development of consensus control theory, there have been mature algorithms and stability analysis methods for consensus control problem in multi-agent systems with precise models.

1.3   MOTIVATION

In this monograph, a series of novel distributed adaptive consensus control methods are presented for nonlinear multi-agent systems from system uncertainties, directed graph, finite-time convergence rate, and event-triggered communication points of view. The state of the art of related research areas and motivation of our work are elaborated from the following three aspects.

1.3.1 ADAPTIVE CONSENSUS CONTROL

In previous section, we reviewed some representative results on consensus control reported in the last two decades. It should be pointed out that most of the aforementioned results can only be applicable to the agents with relative simple or precisely known system dynamics. However, the system uncertainties including unknown parameters, unmodeled system dynamics, and undesired disturbances unavoidably exist in practical applications [56]. Therefore, exploring how to design distributed consensus control schemes for uncertain nonlinear multi-agent systems is of great significance. To this end, some effective control methods have been applied to uncertain nonlinear multi-agent systems, such as robust H_∞ control [68, 69, 101], sliding mode control [103, 158], disturbance observer-based control [23, 36], and so on.

Except for the aforementioned classical control methods, adaptive control techniques are also employed to design distributed consensus control strategies for uncertain multi-agent systems. However, valid distributed adaptive consensus control schemes are still limited, especially for cases with directed communication graphs. This is because directed graph is associated with asymmetric Laplacian matrix, which will bring about new challenges to design distributed adaptive laws based on Lyapunov stability theory as will be discussed in Chapter 2. In [16], a distributed coordination control scheme is proposed for uncertain first-order nonlinear systems by incorporating adaptive neural network with robust control techniques. Bounded synchronization error can be shown if the control gains are selected to be sufficiently large. The results are generalized to second-order and high-order uncertain nonlinear systems in [17] and [161], respectively. Based on the assumption that the common desired trajectory is linearly parameterized with basis functions known by all agents, a distributed tracking control scheme for uncertain first-order nonlinear systems is presented in [155]. Based on the same assumption on the reference trajectory, an adaptive iterative learning control scheme is proposed in [50] for a class of high-order nonlinear multi-agent systems with directed graph. In [128], the derivative of the reference trajectory is assumed to be linearly parameterized. A distributed adaptive tracking control scheme is presented for a class of uncertain first/second multi-agent systems such that asymptotical cooperative tracking can be achieved under undirected graph condition. Apart from these, more results on distributed adaptive consensus control can be referred to [74, 133, 134, 153, 154] and the references therein.

Nevertheless, it should be noted that these results still have some limitations. For example, only bounded consensus convergence performance can be guaranteed; the common desired trajectory takes the linearly parameterized form; the results are established based on undirected graph. To solve these issues, in this monograph, we will address the asymptotic consensus control problem of a class of general uncertain high-order nonlinear multi-agent systems with directed graph.

1.3.2 ADAPTIVE FINITE-TIME CONSENSUS CONTROL

In addition to the intrinsic uncertainties present in multi-agent systems, it is also crucial to address the issue of how to enhance the convergence rate of consensus behavior. This is an important research matter that can contribute to achieving improved overall performance. In the literature, most of the existing consensus control schemes for multi-agent systems pursue asymptotical convergence. This implies that the convergence rate at best is exponential and the consensus errors need an infinite amount of time to converge to the origin.

It is shown in [91] that the second smallest eigenvalue of the Laplacian matrix of the interaction graph determines the consensus convergence rate. To achieve a higher convergence rate, multiple researchers are attempting to identify an improved interaction graph that will result in a larger second smallest eigenvalue. In [52], the problem of finding the best vertex positional configuration is considered such that the second smallest eigenvalue of the associated interaction graph is maximized. In [144], the weights among the agents are designed and the convergence rate is increased by using semi-definite convex programming. However, all these efforts are focused only on choosing proper interaction graphs, rather than finding control schemes to improve performance.

In practice, it is often required that the consensus be reached in finite time. Thus several researchers invoke the finite-time control schemes to guarantee that the consensus can be reached within a specific time frame. In addition, finite-time control schemes possess better disturbance rejection properties as shown in [10]. In [143], a finite-time formation control framework for multi-agent systems of first-order dynamics is developed. In [129], a continuous finite-time control scheme is developed for the state consensus problem for multi-agent systems of first-order dynamics. In [60], the finite-time consensus problem for leaderless and leader-follower multi-agent systems of second-order double integrator dynamics. In [166], the finite-time consensus tracking problem with one leader and the finite-time containment control problem with multiple leaders are considered.

Note that if uncertainties are involved in system dynamics, it is rather difficult to establish exponential convergence by adopting adaptive control tools. The reason is that the use of online parameter estimators that will give rise to a highly nonlinear closed-loop systems, which hinders the potential applications of adaptive control. Therefore ensuring finite-time convergence will be more interesting and significant for adaptive control systems, whereas certain key techniques employed in existing adaptive control literature cannot be applied. For example, Barbalat's lemma normally adopted for analyzing asymptotic convergence cannot be applied to the analysis of finite-time convergence. On the other hand, the finite-time convergence analysis tools adopted in systems without parametric uncertainties, such as those in [9] and [60], cannot be applied to adaptive finite-time control directly. In [40], finite-time stabilization control for a class of single nonlinear system with parametric uncertainties is investigated with a backstepping-like recursive control scheme, but the convergence time is not expressed explicitly. Furthermore, the existing finite-time consensus control schemes mainly focus on first-order integrator as in [129] or second-order double-integrator as in [60]. In this monograph, we will address such an issue for a group of general nonlinear mechanical systems with parametric uncertainties.

1.3.3 EVENT-TRIGGERED ADAPTIVE CONSENSUS CONTROL

In practical applications, multi-agent systems often operate on digital processing platforms. The usual method of information transmission among the agents is through periodic time-triggered communication mechanisms. In order to ensure the stability and satisfactory performance of the closed-loop system, it is often required that there is sufficient communication bandwidth among the connected agents and the communication time interval is sufficiently small. Clearly, this type of communication undoubtedly uses an excessive amount of communication resources, which presents a significant challenge for small agents that have limited resources.

In order to save communication resources of networked systems, the stabilization of nonlinear systems based on event-triggered communication is studied and a basic theoretical framework of event-triggered control is established in [124]. Unlike traditional periodic time-triggered communication mechanisms, communication modules communicate if and only if the pre-designed event-triggered conditions are satisfied. Obviously, such communication mechanism can effectively reduce traffic and save communication resources. Lately, the consensus of multi-agent systems based on event-triggered communication mechanism has been extensively studied by scholars. A comparison of periodic time-triggered communication and event-triggered communication in multi-agent systems is depicted in Figure 1.2. A plenty

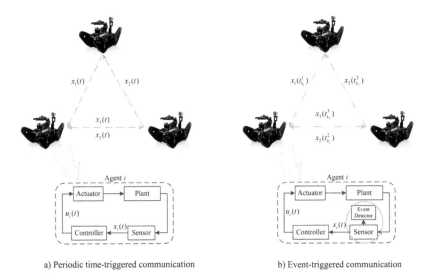

a) Periodic time-triggered communication   b) Event-triggered communication

FIGURE 1.2 Comparison of periodic time-triggered communication and event-triggered communication in multi-agent systems.

of representative works in this area have been reported; see [19, 118, 147, 170] for instance. In [19], a distributed control strategy for average consensus with event-based broadcasting is presented under undirected graph condition. However, Continuous monitoring of the neighbors' states is needed to implement the designed triggering condition. To remove this condition, an exponential decaying triggering threshold is adopted in [118]. In [170], event-based consensus for general linear multi-agent systems under directed graph condition is investigated. The selected triggering functions depend on neighbors' continuous states. Moreover, it is assumed that each agent knows the overall graph which is actually global information. Similar event-based consensus problem is solved in [147] by designing simpler triggering protocols, which are determined only by local state changing rates.

Note that the aforementioned results are mainly established for linear multi-agent systems. However, physical systems are usually nonlinear with system uncertainties in practice. Recently, several adaptive event-triggered consensus control schemes have been proposed for first-order nonlinear systems [137, 159] and second-order nonlinear systems [67,151]. For uncertain high-order nonlinear multi-agent systems, various adaptive backstepping-based consensus control algorithms are developed with continuous communication. Interested readers may refer to [12,74,119,135] and the references therein. However, the event-triggered results are still limited. Under an undirected communication graph, a fuzzy adaptive event-triggered leader-following consensus control algorithm is presented in [61], where continuous monitoring of neighbors' states is required to implement the designed triggering condition. In [131], an observer-based event-triggered adaptive fuzzy control scheme is proposed from the output feedback viewpoint. A triggering condition is elaborately designed to update the controller of each agent. However, the communication among connected subsystems is still continuous. Based on these facts, it can be seen that the consensus control problem for uncertain nonlinear multi-agent systems is not well addressed. In this monograph, we will present a series of novel distributed adaptive consensus control schemes for uncertain nonlinear multi-agent systems with event-triggered communication mechanisms.

1.4 OBJECTIVES

In this monograph, some innovative distributed adaptive consensus control schemes are provided for nonlinear multi-agent systems to handle issues of system uncertainties, directed graph, finite-time convergence rate, and event-triggered communication among connected agents.

In the first part of this monograph, a novel distributed adaptive consensus tracking control scheme will be presented for high-order nonlinear multi-agent systems with unmatched system parameters and directed communication graph will be presented in Chapter 3 under the assumption that the desired trajectory takes the linearly parametric form. In order to relax such an assumption, auxiliary filter-based and hierarchical decomposition-based adaptive consensus tracking control schemes will be further provided in Chapters 4 and 5, respectively.

In the second part of this monograph, a distributed adaptive finite-time consensus control scheme for a class of uncertain nonlinear mechanical systems will be presented in Chapter 6 to improve the consensus convergence rate. With the proposed consensus control scheme, the outputs of all the agents can reach a leaderless consensus within finite time.

In the third part of this monograph, some advances in distributed adaptive consensus control with event-triggered communication mechanisms are presented for the purpose of saving communication resources. First, the leader-following and leaderless consensus control problems for a class of uncertain high-order nonlinear multi-agent systems will be, respectively, discussed in Chapters 7 and 8 based on state feedback control. Then, the consensus tracking control problem for uncertain linear multi-agent systems with unknown system parameters and event-triggered communication will be investigated in Chapter 9 via output feedback control.

In the fourth part of this monograph, the consensus control techniques will be applied to achieve formation control of nonholonomic mobile robots in Chapter 10 and attitude synchronization of multiple spacecraft in Chapter 11, where some novel distributed adaptive cooperative control schemes will be presented.

1.5 PREVIEW OF CHAPTERS

This monograph is composed of 11 chapters. Chapters 2–11 are previewed below.

In Chapter 2, some preliminaries about notations, algebraic graph theory, and adaptive backstepping control technique will be introduced. Besides, the leader-following and leaderless consensus control problems of uncertain first-order nonlinear multi-agent systems under undirected and directed graphs will be investigated, where the main difficulty on distributed adaptive consensus control in directed communication case is pointed out.

In Chapter 3, the consensus tracking control problem for nonlinear multi-agent systems with mismatched uncertainties will be studied under directed graph condition. Suppose that only part of agents can acquire the exact information of the desired trajectory, which takes linearly parametric form. To illustrate the design idea, the first-order multi-agent nonlinear systems are first considered as an example. Then, the obtained results will be extended to the high-order nonlinear multi-agent systems.

In Chapters 4 and 5, the consensus tracking control problem in Chapter 3 is reconsidered. In order to relax the assumption that the desired trajectory takes the linearly parametric form, auxiliary filter-based and hierarchical decomposition-based adaptive consensus tracking control schemes will be presented.

In Chapter 6, we investigate the finite-time leaderless consensus control problem of multi-agent systems consisting of a group of nonlinear mechanical systems with parametric uncertainties. New adaptive finite-time continuous distributed control algorithms will be proposed for the multi-agent systems.

In Chapter 7, we investigate the distributed adaptive consensus tracking problem for a class of uncertain high-order nonlinear multi-agent systems with directed graph and event-triggered communication. For each agent, a group of triggering conditions to broadcast its state information are designed, which are only dependent on its local state changing rate. Then, a totally distributed consensus tracking control scheme based on backstepping technique will be proposed.

In Chapter 8, the same multi-agent systems as in Chapter 7 will be considered. Based on event-triggered communication, a novel distributed adaptive leaderless consensus control law and triggering condition are designed for each agent. Besides, in order to ensure the lower bound of triggering time intervals, a switching triggering condition is presented.

In Chapter 9, an output feedback-based distributed adaptive output consensus tracking control scheme will be presented for heterogenous linear multi-agent systems with unknown system parameters, event-triggered communication, and directed graph.

In Chapter 10, we investigate the formation control problem for multiple nonholonomic mobile robots with unknown system parameters under directed graph condition. The distributed adaptive tracking control strategy in Chapter 3 will be applied to the group of mobile robots to solve the formation control problem.

In Chapter 11, we address the distributed adaptive attitude synchronization problem for multiple rigid spacecraft with unknown inertial matrices and event-triggered communication. Based on a strongly connected directed graph, two distributed event-based adaptive attitude synchronization control schemes are proposed via state and output feedback control.

2 Preliminaries

In this chapter, some preliminaries on distributed adaptive consensus control will be introduced. First, notations used in this monograph and algebraic graph theory are elaborated. Then, adaptive backstepping control technique for uncertain high-order nonlinear systems is reviewed, which will be frequently used in the subsequent chapters. Finally, we respectively investigate the leader-following and leaderless consensus control problems of uncertain first-order nonlinear multi-agent systems under undirected and directed graphs. The design procedure of distributed consensus controller is provided and discussed in detail, where the main difficulty on distributed adaptive consensus control for the directed communication case is pointed out.

2.1 NOTATIONS

In this monograph, $\Re$, $\Re^p$, and $\Re^{p \times q}$ denote the sets of real numbers, $p \times 1$ real vectors, and $p \times q$ real matrices, respectively. Let $\mathbf{1}_p = [1, ..., 1]^T \in \Re^p$ and $\mathbf{0}_p = [0, ..., 0]^T \in \Re^p$. Let $a \in \Re^n$ and $b \in \Re^n$ being two vectors, then define the vector operator $.*$ as $a.* b = [a(1)b(1), ..., a(n)b(n)]^T$. I_p is the $p \times p$ identity matrix. If P is a positive definite matrix, then let $\lambda_{\max}(P)$ and $\lambda_{\min}(P)$ denote its maximum and minimum eigenvalues, respectively. For a vector, $\| \cdot \|$ denotes a standard Euclidean norm. For a matrix, $\| \cdot \|_1$ denotes a standard column sum norm. Let $P = \text{diag}\{P_l\}$ be the diagonal matrix with the lth diagonal entry being P_l. For a function $f(t) : \Re \to \Re$, it is said that $f(t) \in \mathcal{L}_2$ if $(\int_0^\infty f(\tau)^2 d\tau)^{\frac{1}{2}} < \infty$ and $f(t) \in \mathcal{L}_\infty$ if $\sup_{t \geq 0} |f(t)| < \infty$.

2.2 ALGEBRAIC GRAPH THEORY

In this section, we introduce some basic concepts and results about algebraic graph theory. Without loss of generality, a team of N agents is considered in this monograph. To describe the communication or sensing networks among the N agents, a fixed graph $\mathcal{G} \triangleq (\mathcal{A}, \mathcal{V}, \varepsilon)$ is adopted. $\mathcal{A} = [a_{ij}] \in \Re^{N \times N}$ with nonnegative elements is the adjacency matrix associated with a graph $\mathcal{G}$. $\mathcal{V} = \{1, ..., N\}$ and $\varepsilon \subseteq \mathcal{V} \times \mathcal{V}$ denote the set of nodes and edges, respectively. If the edge $(i, j) \in \varepsilon$, then agent j can receive signals from agent i, but not necessarily vice versa [112]. In this case, agent i is called a neighbor of agent j and in turn agent j is called an out-neighbor of agent i. We use $\mathcal{N}_j \triangleq \{i \in \mathcal{V} : (i, j) \in \varepsilon\}$ to denote the collection of neighbors of agent j. The element $a_{ij} = 1$ if $(j, i) \in \varepsilon$ and $a_{ij} = 0$ otherwise. Note that self-edges (i, i) are not allowed. Thus $(i, i) \notin \varepsilon$ and $i \notin \mathcal{N}_i$. As a result, the diagonal elements of $\mathcal{A}$ are all zeros, i.e., $a_{ii} = 0$, $i \in \mathcal{V}$. Let Δ be the diagonal in-degree matrix with diagonal elements being $\Delta_i = \sum_{j \in \mathcal{N}_i} a_{ij}$, $i \in \mathcal{V}$. The Laplacian matrix associated with the adjacency matrix $\mathcal{A}$ is defined as $\mathcal{L} = \Delta - \mathcal{A}$.

DOI: 10.1201/9781003394372-2

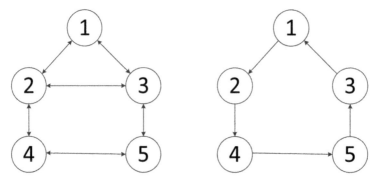

(a) Undirected and connected graph   (b) Weakly connected and balanced graph

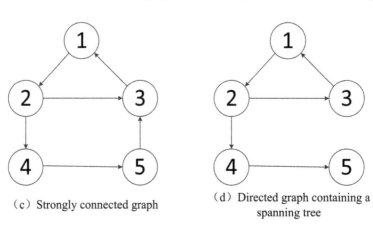

(c) Strongly connected graph

(d) Directed graph containing a spanning tree

FIGURE 2.1   An example of four typical graphs.

In a graph, a node i is balanced if its in-degree equals its out-degree, i.e., $\sum_{j=1}^{N} a_{ij} = \sum_{j=1}^{N} a_{ji}$. A directed path exists, which is originated from node i and terminated at node j if there is a sequence of successive edges $\{(i, k), (k, m), \ldots, (n, l), (l, j)\} \subseteq \varepsilon$ with $i, k, m, \ldots, n, l, j \in \mathcal{V}$.

According to the characteristics of graph connectivity, various kinds of graphs can be defined. Some typical graphs mentioned in this monograph are elaborated in Definition 2.1. For easier understanding, an example of four typical graphs is depicted in Figure 2.1.

Definition 2.1

- **Underlying graph**: *For a digraph, its underlying graph is the graph obtained by replacing all its directed edges with undirected edges.*
- **Weakly connected graph**: *A digraph $\mathcal{G}$ is weakly connected if its underlying graph is connected.*

- **Undirected graph**: *A graph $\mathcal{G}$ is called undirected if there is a connection between agent i and agent j, then $a_{ij} = a_{ji} > 0$; otherwise, $a_{ij} = a_{ji} = 0$.*
- **Undirected and connected graph**: *A graph $\mathcal{G}$ is called undirected and connected if it is an undirected graph and its underlying graph is connected.*
- **Weakly connected and balanced graph**: *A digraph $\mathcal{G}$ is called weakly connected and balanced if its underlying graph is connected and all its nodes are balanced.*
- **Strongly connected graph**: *A digraph $\mathcal{G}$ is called strongly connected if there exists a directed path between any two distinct nodes.*
- **Directed graph containing a spanning tree**: *A digraph $\mathcal{G}$ contains a directed spanning tree if there exists at least a root node i such that all the remaining nodes in the graph can be reached from i through a directed path.*

From the definitions of the Laplacian matrix $\mathcal{L}$ and the graphs, it is straightforward to check that the following properties hold [52, 112].

Property 2.1 *For an undirected graph $\mathcal{G}$, $\mathcal{L}$ is symmetric and positive semi-definite. For a directed graph $\mathcal{G}$, $\mathcal{L}$ is nonsymmetric. In the both cases, 0 is an eigenvalue of $\mathcal{L}$ with an associated eigenvector $\mathbf{1}_N$.*

Property 2.2 *$\mathcal{L}$ is diagonally dominant and has nonnegative diagonal entries. According to Gershgorin's disc theorem, for an undirected graph, all nonzero eigenvalues of $\mathcal{L}$ are real and positive. However, for a directed graph, all nonzero eigenvalues of $\mathcal{L}$ have positive real parts.*

Property 2.3 *$\mathcal{L}$ has a simple zero eigenvalue and all other eigenvalues have positive real parts if and only if $\mathcal{G}$ has a directed spanning tree. Furthermore, if $\mathcal{G}$ is undirected and connected, $\mathcal{L}$ has a simple zero eigenvalue and all other eigenvalues are positive. In addition, $\mathcal{L}\mathbf{1}_N = \mathbf{0}_N$ and there exists a nonnegative vector $\mathbf{p} \in \Re^N$ satisfying $\mathbf{p}^T \mathcal{L} = \mathbf{0}_N$ and $\mathbf{p}^T \mathbf{1}_N = 1$.*

Property 2.4 *For $x = [x_1, \ldots, x_N]^T$, $\mathcal{L}x$ is a column stack vector of $\sum_{j=1}^N a_{ij}(x_i - x_j), i = 1, \ldots, N$. If $\mathcal{G}$ is strongly connected and balanced, then $\mathcal{L}x = \mathbf{0}_N$ or $x^T \mathcal{L}x = \mathbf{0}_N$ if and only if $x_i = x_j$ for $i, j = 1, \ldots, N$.*

Property 2.5 *For an undirected graph, let $\lambda_i(\mathcal{L})$ be the ith eigenvalue of $\mathcal{L}$ with $\lambda_1(\mathcal{L}) \leq \lambda_2(\mathcal{L}) \leq \ldots \leq \lambda_N(\mathcal{L})$, so that $\lambda_1(\mathcal{L}) = 0$. $\lambda_2(\mathcal{L})$ is called* algebraic connectivity, *which is positive if and only if the undirected graph is connected.*

In this monograph, for the consensus tracking case, we use $\mu_i = 1$ to denote the case that agent i can directly access the full information of the desired trajectory $y_r(t)$, otherwise $\mu_i = 0$. Define a pinning matrix as $\mathcal{B} = \text{diag}\{\mu_1, \ldots, \mu_N\}$. The following useful lemmas about the Laplacian matrix $\mathcal{L}$ are introduced, which will be used to analyze the consensus behavior of multi-agent systems in the subsequent chapters.

Lemma 2.1 ([162]) *Let $\xi = [\xi_1, \ldots, \xi_N]^T$ be the left eigenvector of $\mathcal{L}$ associated with eigenvalue $\lambda(\mathcal{L}) = 0$, i.e., $\xi^T \mathcal{L} = \mathbf{0}_N$. For a strongly connected graph, it implies that $\xi_i > 0$, $\sum_{i=1}^{N} \xi_i = 1$. Define $\hat{\mathcal{L}} = (\Omega \mathcal{L} + \mathcal{L}^T \Omega)$ with $\Omega = diag\{\xi_1, \ldots, \xi_N\}$. $\hat{\mathcal{L}}$ is a semi-positive definite matrix with a null space $\mathcal{N}(\hat{\mathcal{L}}) = \{x | x^T \hat{\mathcal{L}} x = 0\} = span\{\mathbf{1}_N\}$.*

Lemma 2.2 ([40, 112]) *If $\mathcal{G}$ has a directed spanning tree and the root node has access to the desired trajectory, then all eigenvalues of $\mathcal{H} = \mathcal{L} + \mathcal{B}$ have positive real parts. If $\mathcal{G}$ is a connected undirected graph and at least one agent has access to the desired trajectory, then $\mathcal{H} = \mathcal{L} + \mathcal{B}$ is symmetric and positive definite.*

Lemma 2.3 ([112]) *Define $Q = (\mathcal{L} + \mathcal{B}) + (\mathcal{L} + \mathcal{B})^T$. If $\mathcal{G}$ is a weakly connected and balanced graph, the matrix Q is symmetric positive definite.*

Lemma 2.4 ([66, 163]) *If the communication graph has a directed spanning tree with the root node having access to the desired trajectory, then the matrix $\mathcal{L} + \mathcal{B}$ is nonsingular where $\mathcal{B} = diag\{\mu_1, \ldots, \mu_N\}$. Define*

$$p = [P_1, \ldots, P_N]^T = (\mathcal{L} + \mathcal{B})^{-T} [1, \ldots, 1]^T$$
$$P = diag\{P_1, \ldots, P_N\}$$
$$Q = P(\mathcal{L} + \mathcal{B}) + (\mathcal{L} + \mathcal{B})^T P,$$

then $P > 0$ and $Q > 0$.

2.3   ADAPTIVE BACKSTEPPING CONTROL

In the early 1990s, a novel approach known as Backstepping [56] was proposed for the design of adaptive controllers. Backstepping offers a recursive Lyapunov-based framework specifically tailored to strict feedback systems. Notably, when the controlled plant can be transformed into the parametric-strict feedback form, this approach ensures system stabilization and certain specific output regulation properties. One notable strength of the backstepping design method lies in its systematic step-by-step procedure for designing stabilizing controllers. In those immediate steps, some state variables are selected as virtual controls and stabilizing functions are designed correspondingly.

To handle systems with parametric uncertainties, adaptive backstepping controllers are designed by incorporating the estimated parameters. Similar to conventional adaptive control methods, adaptive backstepping control systems can be constructed through direct or indirect approaches [47]. In direct adaptive backstepping control, parameter estimators are designed together with controllers based on the Lyapunov functions augmented by the squared terms of parameter estimation errors. By leveraging the tuning function technique, the challenge of over-parametrization can be mitigated, leading to a reduction in the implementation cost of the adaptive control scheme.

In this section, adaptive backstepping control technique will be introduced. The procedures to design state feedback adaptive controllers by incorporating the tuning functions are then presented. In the second part, the output feedback control scheme is proposed for a class of parametric strict-feedback nonlinear systems and stability analysis is also provided.

2.3.1   STATE FEEDBACK CONTROL

We consider a class of nonlinear systems as follows,

$$\dot{x}_1 = x_2 + \varphi_1^T(x_1)\theta$$
$$\dot{x}_2 = x_3 + \varphi_2^T(x_1, x_2)\theta$$
$$\vdots \quad \vdots$$
$$\dot{x}_{n-1} = x_n + \varphi_{n-1}^T(x_1, \dots, x_{n-1})\theta$$
$$\dot{x}_n = \varphi_0(x) + \varphi_n^T(x)\theta + \beta(x)u$$
$$y = x_1, \tag{2.1}$$

where $x = [x_1, \dots, x_n]^T \in \Re^n$, $u \in \Re$, and $y \in \Re$ are the state, input and output of the system, respectively. $\theta \in \Re^p$ is an unknown constant vector, $\varphi_0 \in \Re$, $\varphi_i \in \Re^p$ for $i = 1, \dots, n$, β are known smooth nonlinear functions. Note that the class of nonlinear systems in the form of (2.1) is known as parametric strict-feedback systems since there are only feedback paths except for the integrators and the nonlinearities depend only on variables which are "feed back" [56].

The control objective is to force the system output to asymptotically track a reference signal $y_r(t)$ while ensuring system stability. To achieve the objective, the following assumptions are imposed.

Assumption 2.3.1 *The reference signal $y_r(t)$ and its first n derivatives $y_r^{(i)}$, $i = 1, \dots, n$ are known, bounded, and piecewise continuous.*

Assumption 2.3.2 $\beta(x) \neq 0, \forall x \in \Re^n$.

Here, n steps are required to determine the control signal for the n-th order system in (2.1). The design procedure is elaborated as follows.

Step 1. We introduce the first two error variables in this step

$$z_1 = y - y_r \tag{2.2}$$
$$z_2 = x_2 - \dot{y}_r - \alpha_1, \tag{2.3}$$

where z_1 implies the tracking error, of which the convergence $\lim\limits_{t\to\infty} z_1(t) = 0$ is to be achieved. The z_1 dynamics is derived as

$$\dot{z}_1 = \dot{y} - \dot{y}_r$$

$$\begin{aligned}
&= x_2 + \varphi_1^T \theta - \dot{y}_r \\
&= z_2 + \alpha_1 + \varphi_1^T \theta.
\end{aligned} \tag{2.4}$$

α_1 is the first stabilizing function designed as

$$\alpha_1 = -c_1 z_1 - \varphi_1^T \hat{\theta}, \tag{2.5}$$

where c_1 is a positive constant and $\hat{\theta}$ is an estimate of θ. In fact, α_1 is the "desired value" of x_2 to stabilize $\dot{z}_1$ system as seen from the second equation of (2.4) if $\dot{y}_r = 0$. Thus z_2 is the error between the actual and "desired" values of x_2 augmented by the term $-\dot{y}_r$.

Then, a Lyapunov function is defined at this step.

$$V_1 = \frac{1}{2} z_1^2 + \frac{1}{2} \tilde{\theta}^T \Gamma^{-1} \tilde{\theta}, \tag{2.6}$$

where Γ is a positive definite matrix and $\tilde{\theta}$ is the estimation error that $\tilde{\theta} = \theta - \hat{\theta}$. From (2.4) and (2.5), the derivative of V_1 is derived as

$$\begin{aligned}
\dot{V}_1 &= z_1 \left(-c_1 z_1 + z_2 + \varphi_1^T \tilde{\theta} \right) - \tilde{\theta}^T \Gamma^{-1} \dot{\hat{\theta}} \\
&= z_1 (-c_1 z_1 + z_2) - \tilde{\theta}^T \left(\Gamma^{-1} \dot{\hat{\theta}} - \varphi_1 z_1 \right).
\end{aligned} \tag{2.7}$$

We define the first tuning function as

$$\tau_1 = \varphi_1 z_1. \tag{2.8}$$

Substituting (2.8) into (2.7), we obtain that

$$\dot{V}_1 = -c_1 z_1^2 + z_1 z_2 - \tilde{\theta}^T \left(\Gamma^{-1} \dot{\hat{\theta}} - \tau_1 \right). \tag{2.9}$$

Step 2. We now treat the second equation of (2.1) by considering x_3 as the control variable. Introduce an error variable

$$z_3 = x_3 - \ddot{y}_r - \alpha_2. \tag{2.10}$$

Taking the derivative of z_2, we have

$$\begin{aligned}
\dot{z}_2 &= \dot{x}_2 - \ddot{y}_r - \dot{\alpha}_1 \\
&= z_3 + \alpha_2 - \frac{\partial \alpha_1}{\partial x_1} x_2 + \left(\varphi_2 - \frac{\partial \alpha_1}{\partial x_1} \varphi_1 \right)^T \theta - \frac{\partial \alpha_1}{\partial y_r} \dot{y}_r - \frac{\partial \alpha_1}{\partial \hat{\theta}} \dot{\hat{\theta}},
\end{aligned} \tag{2.11}$$

where the fact that α_1 is a function of x_1, y_r, and $\hat{\theta}$ has been utilized. α_2 is the second stabilizing function designed at this step to stabilize (z_1, z_2)-system composed of (2.4) and (2.11). We select α_2 as

$$\alpha_2 = -z_1 - c_2 z_2 + \frac{\partial \alpha_1}{\partial x_1} x_2 - \left(\varphi_2 - \frac{\partial \alpha_1}{\partial x_1} \varphi_1 \right)^T \hat{\theta} + \frac{\partial \alpha_1}{\partial y_r} \dot{y}_r + \frac{\partial \alpha_1}{\partial \hat{\theta}} \Gamma \tau_2, \tag{2.12}$$

where c_2 is a positive constant and τ_2 is the second tuning function designed based on τ_1 such that

$$\tau_2 = \tau_1 + \left(\varphi_2 - \frac{\partial \alpha_1}{\partial x_1} \varphi_1 \right) z_2. \tag{2.13}$$

We now define a Lyapunov function V_2 as

$$V_2 = V_1 + \frac{1}{2} z_2^2. \tag{2.14}$$

From (2.9), (2.11)–(2.13), the derivative of V_2 is computed as

$$
\begin{aligned}
\dot{V}_2 = & - c_1 z_1^2 + z_1 z_2 - \tilde{\theta}^T \left(\Gamma^{-1} \dot{\hat{\theta}} - \tau_1 \right) + z_2(-z_1 - c_2 z_2 + z_3) \\
& + z_2 \left(\varphi_2 - \frac{\partial \alpha_1}{\partial x_1} \varphi_1 \right)^T \tilde{\theta} + z_2 \frac{\partial \alpha_1}{\partial \hat{\theta}} \left(\Gamma \tau_2 - \dot{\hat{\theta}} \right) \\
= & - c_1 z_1^2 - c_2 z_2^2 + z_2 z_3 + \tilde{\theta}^T \left(\tau_2 - \Gamma^{-1} \dot{\hat{\theta}} \right) + z_2 \frac{\partial \alpha_1}{\partial \hat{\theta}} \left(\Gamma \tau_2 - \dot{\hat{\theta}} \right). \tag{2.15}
\end{aligned}
$$

Note that if x_3 were the actual control, we have $z_3 = 0$. If the parameter update law were chosen as $\dot{\hat{\theta}} = \Gamma \tau_2$, $\dot{V}_2 = -c_1 z_1^2 - c_2 z_2^2$ is rendered negative semi-definite and the (z_1, z_2)-system can be stabilized. However, x_3 is not the actual control. Similar to $z_1 z_2$ canceled at this step, the term $z_2 z_3$ will be canceled at the next step. Moreover, the discrepancy between $\Gamma \tau_2$ and $\dot{\hat{\theta}}$ will be compensated partly by defining another tuning function τ_3 at the next step.

Step 3. We proceed to treat the third equation of (2.1). Introduce that

$$z_4 = x_4 - y_r^{(3)} - \alpha_3. \tag{2.16}$$

Computing the derivative of z_3, we have

$$
\begin{aligned}
\dot{z}_3 = & z_4 + \alpha_3 - \frac{\partial \alpha_2}{\partial x_1} x_2 - \frac{\partial \alpha_2}{\partial x_2} x_3 + \left(\varphi_3 - \frac{\partial \alpha_2}{\partial x_1} \varphi_1 - \frac{\partial \alpha_2}{\partial x_2} \varphi_2 \right)^T \theta - \frac{\partial \alpha_2}{\partial y_r} \dot{y}_r \\
& - \frac{\partial \alpha_2}{\partial \dot{y}_r} \ddot{y}_r - \frac{\partial \alpha_2}{\partial \hat{\theta}} \dot{\hat{\theta}}, \tag{2.17}
\end{aligned}
$$

where the fact that α_2 is a function of x_1, x_2, y_r, $\dot{y}_r$ and $\hat{\theta}$ has been utilized. We then select α_3 as

$$
\begin{aligned}
\alpha_3 = & - z_2 - c_3 z_3 + \frac{\partial \alpha_2}{\partial x_1} x_2 + \frac{\partial \alpha_2}{\partial x_2} x_3 - \left(\varphi_3 - \frac{\partial \alpha_2}{\partial x_1} \varphi_1 - \frac{\partial \alpha_2}{\partial x_2} \varphi_2 \right)^T \hat{\theta} \\
& + \frac{\partial \alpha_2}{\partial y_r} \dot{y}_r + \frac{\partial \alpha_2}{\partial \dot{y}_r} \ddot{y}_r + \frac{\partial \alpha_2}{\partial \hat{\theta}} \Gamma \tau_3 + z_2 \frac{\partial \alpha_1}{\partial \hat{\theta}} \Gamma \left(\varphi_3 - \frac{\partial \alpha_2}{\partial x_1} \varphi_1 - \frac{\partial \alpha_2}{\partial x_2} \varphi_2 \right), \tag{2.18}
\end{aligned}
$$

where c_3 is a positive constant and τ_3 is the third tuning function designed based on τ_2 so that

$$\tau_3 = \tau_2 + \left(\varphi_3 - \frac{\partial \alpha_2}{\partial x_1}\varphi_1 - \frac{\partial \alpha_2}{\partial x_2}\varphi_2 \right) z_3. \qquad (2.19)$$

The (z_1, z_2, z_3)-system (2.4), (2.11), and (2.17) is stabilized with respect to the Lyapunov function

$$V_3 = V_2 + \frac{1}{2}z_3^2, \qquad (2.20)$$

whose derivative is

$$\dot{V}_3 = - c_1 z_1^2 - c_2 z_2^2 - c_3 z_3^2 + z_3 z_4 + \tilde{\theta}^T \left(\tau_3 - \Gamma^{-1}\dot{\hat{\theta}} \right) + z_2 \frac{\partial \alpha_1}{\partial \hat{\theta}} \left(\Gamma \tau_2 - \dot{\hat{\theta}} \right)$$

$$+ z_3 \frac{\partial \alpha_2}{\partial \hat{\theta}} \left(\Gamma \tau_3 - \dot{\hat{\theta}} \right) + z_2 \frac{\partial \alpha_1}{\partial \hat{\theta}} \Gamma \left(\varphi_3 - \frac{\partial \alpha_2}{\partial x_1}\varphi_1 - \frac{\partial \alpha_2}{\partial x_2}\varphi_2 \right) z_3. \quad (2.21)$$

Note that

$$z_2 \frac{\partial \alpha_1}{\partial \hat{\theta}} \left(\Gamma \tau_2 - \dot{\hat{\theta}} \right) = z_2 \frac{\partial \alpha_1}{\partial \hat{\theta}} \left(\Gamma \tau_3 - \dot{\hat{\theta}} \right) + z_2 \frac{\partial \alpha_1}{\partial \hat{\theta}}(\Gamma \tau_2 - \Gamma \tau_3)$$

$$= z_2 \frac{\partial \alpha_1}{\partial \hat{\theta}} \left(\Gamma \tau_3 - \dot{\hat{\theta}} \right) - z_2 \frac{\partial \alpha_1}{\partial \hat{\theta}} \Gamma \left(\varphi_3 - \frac{\partial \alpha_2}{\partial x_1}\varphi_1 - \frac{\partial \alpha_2}{\partial x_2}\varphi_2 \right)$$

$$\times z_3. \qquad (2.22)$$

Substituting (2.22) into (2.21), we obtain

$$\dot{V}_3 = - c_1 z_1^2 - c_2 z_2^2 - c_3 z_3^2 + z_3 z_4 + \tilde{\theta}^T \left(\tau_3 - \Gamma^{-1}\dot{\hat{\theta}} \right) + \left(z_2 \frac{\partial \alpha_1}{\partial \hat{\theta}} + z_3 \frac{\partial \alpha_2}{\partial \hat{\theta}} \right)$$

$$\times \left(\Gamma \tau_3 - \dot{\hat{\theta}} \right). \qquad (2.23)$$

From the discussion above, we can see that the last term of the designed α_3 in (2.18) is important to cancel the term $z_2 \frac{\partial \alpha_1}{\partial \hat{\theta}}(\Gamma \tau_2 - \Gamma \tau_3)$ in rewriting the term $z_2 \frac{\partial \alpha_1}{\partial \hat{\theta}}(\Gamma \tau_2 - \dot{\hat{\theta}})$ as in (2.22).

Step i, $(i = 4, \ldots, n-1)$. Introduce the error variable

$$z_i = x_i - y_r^{(i-1)} - \alpha_{i-1} \qquad (2.24)$$

The dynamics of z_i is derived as

$$\dot{z}_i = z_{i+1} + \alpha_i - \sum_{k=1}^{i-1} \frac{\partial \alpha_{i-1}}{\partial x_k}x_{k+1} + \left(\varphi_i - \sum_{k=1}^{i-1} \frac{\partial \alpha_{i-1}}{\partial x_k}\varphi_k \right)^T \theta$$

$$- \sum_{k=1}^{i-1} \frac{\partial \alpha_{i-1}}{\partial y_r^{(k-1)}}y_r^{(k)} - \frac{\partial \alpha_{i-1}}{\partial \hat{\theta}}\dot{\hat{\theta}}. \qquad (2.25)$$

The stabilization function α_i is chosen as

$$
\alpha_i = -z_{i-1} - c_i z_i + \sum_{k=1}^{i-1} \frac{\partial \alpha_{i-1}}{\partial x_k} x_{k+1} - \left(\varphi_i - \sum_{k=1}^{i-1} \frac{\partial \alpha_{i-1}}{\partial x_k} \varphi_k \right)^T \hat{\theta}
$$

$$
+ \sum_{k=1}^{i-1} \frac{\partial \alpha_{i-1}}{\partial y_r^{(k-1)}} y_r^{(k)} + \frac{\partial \alpha_{i-1}}{\partial \hat{\theta}} \Gamma \tau_i + \sum_{k=2}^{i-1} z_k \frac{\partial \alpha_{k-1}}{\partial \hat{\theta}} \Gamma
$$

$$
\times \left(\varphi_i - \sum_{j=1}^{i-1} \frac{\partial \alpha_{i-1}}{\partial x_j} \varphi_j \right), \tag{2.26}
$$

where c_i is a positive constant and τ_i is the ith tuning function defined as

$$
\tau_i = \tau_{i-1} + \left(\varphi_i - \sum_{k=1}^{i-1} \frac{\partial \alpha_{i-1}}{\partial x_k} \varphi_k \right) z_i. \tag{2.27}
$$

The $(z_1, \ldots, z_i)$-system is stabilized with respect to the Lyapunov function defined as

$$
V_i = V_{i-1} + \frac{1}{2} z_i^2, \tag{2.28}
$$

whose derivative is

$$
\dot{V}_i = -\sum_{k=1}^{i} c_k z_k^2 + z_i z_{i+1} + \tilde{\theta}^T \left(\tau_i - \Gamma^{-1} \dot{\hat{\theta}} \right) + \left(\sum_{k=2}^{i} z_k \frac{\partial \alpha_{k-1}}{\partial \hat{\theta}} \right)
$$

$$
\times \left(\Gamma \tau_i - \dot{\hat{\theta}} \right). \tag{2.29}
$$

Step n We introduce

$$
z_n = x_n - y_r^{(n-1)} - \alpha_{n-1}. \tag{2.30}
$$

The derivative of z_n is

$$
\dot{z}_n = \varphi_0 + \beta u - \sum_{k=1}^{n-1} \frac{\partial \alpha_{n-1}}{\partial x_k} x_{k+1} + \left(\varphi_n - \sum_{k=1}^{n-1} \frac{\partial \alpha_{n-1}}{\partial x_k} \varphi_k \right)^T \theta
$$

$$
- \sum_{k=1}^{n-1} \frac{\partial \alpha_{n-1}}{\partial y_r^{(k-1)}} y_r^{(k)} - y_r^{(n)} - \frac{\partial \alpha_{n-1}}{\partial \hat{\theta}} \dot{\hat{\theta}}. \tag{2.31}
$$

The control input u is designed as

$$
u = \frac{1}{\beta} \left[\alpha_n + y_r^{(n)} \right], \tag{2.32}
$$

with

$$
\begin{aligned}
\alpha_n = & -z_{n-1} - c_n z_n - \varphi_0 + \sum_{k=1}^{n-1} \frac{\partial \alpha_{n-1}}{\partial x_k} x_{k+1} - \left(\varphi_n - \sum_{k=1}^{n-1} \frac{\partial \alpha_{n-1}}{\partial x_k} \varphi_k \right)^T \hat{\theta} \\
& + \sum_{k=1}^{n-1} \frac{\partial \alpha_{n-1}}{\partial y_r^{(k-1)}} y_r^{(k)} + \frac{\partial \alpha_{n-1}}{\partial \hat{\theta}} \Gamma \tau_n + \sum_{k=2}^{n-1} z_k \frac{\partial \alpha_{k-1}}{\partial \hat{\theta}} \Gamma \\
& \times \left(\varphi_n - \sum_{j=1}^{n-1} \frac{\partial \alpha_{n-1}}{\partial x_j} \varphi_j \right),
\end{aligned}
\tag{2.33}
$$

where c_n is a positive constant and τ_n is

$$
\tau_n = \tau_{n-1} + \left(\varphi_n - \sum_{k=1}^{n-1} \frac{\partial \alpha_{n-1}}{\partial x_k} \varphi_k \right).
\tag{2.34}
$$

Define the Lyapunov function as

$$
V_n = V_{n-1} + \frac{1}{2} z_n^2,
\tag{2.35}
$$

whose derivative is computed as

$$
\dot{V}_n = -\sum_{k=1}^{n} c_k z_k^2 + \tilde{\theta}^T \left(\tau_n - \Gamma^{-1} \dot{\hat{\theta}} \right) + \left(\sum_{k=2}^{n} z_k \frac{\partial \alpha_{k-1}}{\partial \hat{\theta}} \right) \left(\Gamma \tau_n - \dot{\hat{\theta}} \right).
\tag{2.36}
$$

By determining the parameter update law as

$$
\dot{\hat{\theta}} = \Gamma \tau_n,
\tag{2.37}
$$

$\dot{V}_n$ is rendered negative definite that

$$
\dot{V}_n = -\sum_{k=1}^{n} c_k z_k^2.
\tag{2.38}
$$

From the definition of V_n and (2.38), it follows that $z_1, \cdots, z_n, \tilde{\theta}$ are bounded. Since $\hat{\theta} = \theta - \tilde{\theta}$, $\hat{\theta}$ is also bounded. From (2.2) and Assumption 2.3.1, y is bounded. From (2.5) and smoothness of $\varphi_1(x_1)$, α_1 is bounded. Combining with the definition of z_2 in (2.3) and the boundedness of $\dot{y}_r$, it follows that x_2 is bounded. By following similar procedure, the boundedness of α_i for $i = 2, \ldots, n$, x_i for $i = 3, \ldots, n$ is also ensured. From (2.32), we can conclude that the control signal u is bounded. Thus, the boundedness of all the signals in the closed-loop adaptive system is guaranteed. Furthermore, we define $z = [z_1, \ldots, z_n]^T$. From the LaSalle-Yoshizawa Theorem, $\lim_{t \to \infty} z(t) = 0$. This implies that asymptotic tracking is also achieved, i.e., $\lim_{t \to \infty} [y(t) - y_r(t)] = 0$. The above facts are formally stated in the following theorem.

Theorem 2.1 *Consider the plant (2.1) under Assumptions 2.3.1–2.3.2. The controller (2.32) and the parameter update law (2.37) guarantee the global boundedness of all signals in the closed-loop adaptive system and the asymptotic tracking is achieved, i.e.,* $\lim_{t \to \infty} [y(t) - y_r(t)] = 0$.

2.3.2 OUTPUT FEEDBACK CONTROL

Now backstepping design procedures with output feedback for nonlinear systems described in the following form are introduced.

$$\dot{x}_1 = x_2 + \varphi_1^T(y)\theta$$

$$\vdots$$

$$\dot{x}_{\rho-1} = x_\rho + \varphi_{\rho-1}^T(y)\theta$$

$$\dot{x}_\rho = x_{\rho+1} + \varphi_\rho^T(y)\theta + b_m u$$

$$\vdots$$

$$\dot{x}_{n-1} = x_n + \varphi_{n-1}^T(y)\theta + b_1 u$$

$$\dot{x}_n = \varphi_n^T(y)\theta + b_0 u$$

$$y = x_1 \tag{2.39}$$

where $x = [x_1, \ldots, x_n]^T \in \Re^n$, $u \in \Re$, and $y \in \Re$ are the state, input, and output of the system, respectively. $\theta \in \Re^p$ is an unknown constant vector, $\varphi_0 \in \Re$, $\varphi_i \in \Re^p$ for $i = 1, \ldots, n$, β are known smooth nonlinear functions, and $b_m, \ldots, b_0$ are unknown constants.

The control objective is to force the system output to asymptotically track a reference signal $y_r(t)$ while ensuring system stability. To achieve the objective, the following assumptions are imposed.

Assumption 2.3.3 *The sign of b_m is known.*

Assumption 2.3.4 *The relative degree $\rho = n - m$ is known and the system is minimum phase.*

Assumption 2.3.5 *The reference signal y_r and its ρ-th order derivatives are piecewise continuous, known and bounded.*

2.3.2.1 State Estimation Filters

In order to design the desired adaptive output feedback control law, we rewrite the system (2.39) in the following form

$$\dot{x} = Ax + \Phi(y)\theta + \begin{bmatrix} 0 \\ b \end{bmatrix} u \tag{2.40}$$

where

$$
A = \begin{bmatrix} 0 & 1 & 0 & \cdots & 0 \\ 0 & 0 & 1 & \cdots & 0 \\ \vdots & \vdots & \vdots & \ddots & \vdots \\ 0 & 0 & 0 & \cdots & 1 \\ 0 & 0 & 0 & \cdots & 0 \end{bmatrix}, \Phi(y) = \begin{bmatrix} \varphi_1^T(y) \\ \vdots \\ \varphi_n^T(y) \end{bmatrix}, b = \begin{bmatrix} b_m \\ \vdots \\ b_0 \end{bmatrix} \qquad (2.41)
$$

Note that only output y is measured. Thus x is unavailable. We need to design filters to estimate x and generate some signals for controller design. These filters are summarized as

$$
\dot{\xi} = A_0 \xi + ky \qquad (2.42)
$$

$$
\dot{\Xi}^T = A_0 \Xi^T + \Phi(y) \qquad (2.43)
$$

$$
\dot{\lambda} = A_0 \lambda + e_n u \qquad (2.44)
$$

$$
v_i = A_0^i \lambda, \quad i = 0, 1, \ldots, m \qquad (2.45)
$$

where $k = [k_1, \ldots, k_n]^T$ such that all eigenvalues of $A_0 = A - k e_1^T$ are at some desired stable locations. The state estimates are given by

$$
\hat{x}(t) = \xi + \Xi^T \theta + \sum_{i=0}^{m} b_i v_i \qquad (2.46)
$$

Note that $\hat{x}$ is unavailable due to the unknown parameters θ and b, so the estimate $\hat{x}$ cannot be used in the later controller design. Instead, it will be used for stability analysis. The derivative of $\hat{x}$ is given as

$$
\begin{aligned}
\dot{\hat{x}}(t) &= \dot{\xi} + \dot{\Xi}^T \theta + \sum_{i=0}^{m} b_i \dot{v}_i \\
&= A_0 \xi + ky + \left(A_0 \Xi^T + \Phi(y) \right) \theta + \sum_{i=0}^{m} b_i A_0^i \left(A_0 \lambda + e_n u \right) \\
&= A_0 \left(\xi + \Xi^T \theta + \sum_{i=0}^{m} b_i v_i \right) + ky + \Phi(y)\theta + \begin{bmatrix} 0 \\ b \end{bmatrix} u \\
&= A_0 \hat{x} + ky + \Phi(y)\theta + \begin{bmatrix} 0 \\ b \end{bmatrix} u \qquad (2.47)
\end{aligned}
$$

It can be shown that the state estimation error

$$
\epsilon = x(t) - \hat{x}(t) \qquad (2.48)
$$

satisfies

$$\begin{aligned}
\dot{\epsilon} &= \dot{x}(t) - \dot{\hat{x}}(t) \\
&= Ax - ky - A_0\hat{x} \\
&= \left(A_0 + ke_1^T\right)x - ky - A_0\hat{x} \\
&= A_0\epsilon
\end{aligned}$$
(2.49)

Suppose $P \in \Re^{n \times n}$ is a positive definite matrix, satisfying $PA_0 + A_0^T P \leq -I$ and let

$$V_\epsilon = \epsilon^T P \epsilon$$
(2.50)

It can be shown that

$$\dot{V}_\epsilon = \epsilon^T \left(PA_0 + A_0^T P\right)\epsilon$$
(2.51)

$$\leq -\epsilon^T \epsilon$$
(2.52)

This Lyapunov function guarantees that $\epsilon \to 0$, which implies $\hat{x}(t) \to x(t)$. Note that the backstepping design starts with its output y, which is the only available system state allowed to appear in the control law. The dynamic equation of y is expressed as

$$\begin{aligned}
\dot{y} &= x_2 + \varphi_1^T(y)\theta \\
&= b_m v_{m,2} + \xi_2 + \bar{\omega}^T \Theta + \epsilon_2
\end{aligned}$$
(2.53)

where

$$\Theta = \left[b_m, \dots, b_0, \theta^T\right]^T$$
(2.54)

$$\omega = \left[v_{m,2}, v_{m-1,2}, \dots, v_{0,2}, \Xi_2 + \varphi_1^T\right]^T$$
(2.55)

$$\bar{\omega} = \left[0, v_{m-1,2}, \dots, v_{0,2}, \Xi_2 + \varphi_1^T\right]^T$$
(2.56)

In above equations, $\epsilon_2, v_{i,2}, \xi_2$, and Ξ_2 denote the second entries of ϵ, v_i, ξ, and Ξ, respectively, and y, v_i, ξ, Ξ are all available signals.

Combining system (2.53) with our filters (2.42)–(2.45), system (2.39) is represented as

$$\dot{y} = b_m v_{m,2} + \xi_2 + \bar{\omega}^T \Theta + \epsilon_2$$
(2.57)

$$\dot{v}_{m,i} = v_{m,i+1} - k_i v_{m,1}, \quad i = 2, 3, \dots, \rho - 1$$
(2.58)

$$\dot{v}_{m,\rho} = v_{m,\rho+1} - k_\rho v_{m,1} + u$$
(2.59)

System (2.57)–(2.59) will be our design system, whose states $y, v_{m,2}, \dots, v_{m,\rho}$ are available. Our task at this stage is to globally stabilize the system and also to achieve the asymptotic tracking of y_r by y.

2.3.2.2 Design Procedure and Stability Analysis

In this section, we present the adaptive control design using the backstepping technique with tuning functions in ρ steps. Firstly, we take the change of coordinates

$$z_1 = y - y_r \tag{2.60}$$

$$z_i = v_{m,i} - \alpha_{i-1} - \hat{\varrho} y_r^{(i-1)}, \quad i = 2, 3, \dots, \rho, \tag{2.61}$$

where $\hat{\varrho}$ is an estimate of $\varrho = 1/b_m$, α_{i-1} is the virtual control at each step and will be determined in later discussions.

Step 1. Starting with the equation for the tracking error z_1, we obtain, from (2.57) and (2.60), that

$$\dot{z}_1 = b_m v_{m,2} + \xi_2 + \bar{\omega}^T \Theta + \epsilon_2 - \dot{y}_r \tag{2.62}$$

By substituting (2.61) for $i = 2$ into (2.62) and using $\tilde{\varrho} = \frac{1}{b_m} - \hat{\varrho}$, we get

$$\dot{z}_1 = b_m \alpha_1 + \xi_2 + \bar{\omega}^T \Theta + \epsilon_2 - b_m \tilde{\varrho} \dot{y}_r + b_m z_2 \tag{2.63}$$

By considering $v_{m,2}$ as the first virtual control, we select a virtual control law α_1 as

$$\alpha_1 = \hat{\varrho} \bar{\alpha}_1 \tag{2.64}$$

$$\bar{\alpha}_1 = - c_1 z_1 - d_1 z_1 - \xi_2 - \bar{\omega}^T \hat{\Theta} \tag{2.65}$$

where c_1 and d_1 are positive design parameters, and $\hat{\Theta}$ is the estimate of Θ. From (2.63) and (2.64) we have

$$\begin{aligned}
\dot{z}_1 &= - c_1 z_1 - d_1 z_1 + \epsilon_2 + \bar{\omega}^T \tilde{\Theta} - b_m \left(\dot{y}_r + \bar{\alpha}_1 \right) \tilde{\varrho} + b_m z_2 \\
&= - \left(c_1 + d_1 \right) z_1 + \epsilon_2 + \left(\omega - \hat{\varrho} \left(\dot{y}_r + \bar{\alpha}_1 \right) e_1 \right)^T \tilde{\Theta} \\
&\quad - b_m \left(\dot{y}_r + \bar{\alpha}_1 \right) \tilde{\varrho} + \hat{b}_m z_2
\end{aligned} \tag{2.66}$$

where $\tilde{\Theta} = \Theta - \hat{\Theta}$. Note that

$$b_m \alpha_1 = b_m \hat{\varrho} \bar{\alpha}_1 = \bar{\alpha}_1 - b_m \tilde{\varrho} \bar{\alpha}_1 \tag{2.67}$$

$$\begin{aligned}
\bar{\omega}^T \tilde{\Theta} + b_m z_2 &= \bar{\omega}^T \tilde{\Theta} + \hat{b}_m z_2 + \tilde{b}_m z_2 \\
&= \bar{\omega}^T \tilde{\Theta} + \left(v_{m,2} - \hat{\varrho} \dot{y}_r - \alpha_1 \right) e_1^T \tilde{\Theta} + \hat{b}_m z_2 \\
&= \left(\omega - \hat{\varrho} \left(\dot{y}_r + \bar{\alpha}_1 \right) e_1 \right)^T \tilde{\Theta} + \hat{b}_m z_2
\end{aligned} \tag{2.68}$$

Define the Lyapunov function V_1 as

$$V_1 = \frac{1}{2}z_1^2 + \frac{1}{2}\tilde{\Theta}^T\Gamma^{-1}\tilde{\Theta} + \frac{|b_m|}{2\gamma}\tilde{\varrho}^2 + \frac{1}{2d_1}\epsilon^T P\epsilon \tag{2.69}$$

where Γ is a positive definite design matrix, γ is a positive design parameter, and P is a definite positive matrix such that $PA_0 + A_0^T P = -I, P = P^T > 0$. We examine the derivative of V_1

$$\dot{V}_1 \leq z_1\dot{z}_1 - \tilde{\Theta}^T\Gamma^{-1}\dot{\hat{\Theta}} - \frac{|b_m|}{\gamma}\tilde{\varrho}\dot{\hat{\varrho}} - \frac{1}{2d_1}\epsilon^T\epsilon$$

$$\leq -c_1z_1^2 + \hat{b}_m z_1 z_2 - \frac{1}{4d_1}\epsilon^T\epsilon - d_1 z_1^2 + z_1\epsilon_2 - \frac{\|\epsilon\|^2}{4d_1}$$

$$- |b_m|\tilde{\varrho}\frac{1}{\gamma}\left[\gamma\,\text{sign}\,(b_m)\,(\dot{y}_r + \bar{\alpha}_1)\,z_1 + \dot{\hat{\varrho}}\right]$$

$$+ \tilde{\Theta}^T\left[(\omega - \hat{\varrho}\,(\dot{y}_r + \bar{\alpha}_1)\,e_1)\,z_1 - \Gamma^{-1}\dot{\hat{\Theta}}\right] \tag{2.70}$$

Now we choose

$$\dot{\hat{\varrho}} = -\gamma\,\text{sign}\,(b_m)\,(\dot{y}_r + \bar{\alpha}_1)\,z_1 \tag{2.71}$$

Define

$$\tau_1 = (\omega - \hat{\varrho}\,(\dot{y}_r + \bar{\alpha}_1)\,e_1)\,z_1 \tag{2.72}$$

and τ_1 is called the first tuning function. Then the following can be derived by using Young's inequality $ab \leq d_1 a^2 + \frac{1}{4d_1}b^2$, update law (2.71) and (2.72)

$$\dot{V}_1 \leq -c_1 z_1^2 + \hat{b}_m z_1 z_2 - \frac{1}{4d_1}\epsilon^T\epsilon + \tilde{\Theta}^T\left(\tau_1 - \Gamma^{-1}\dot{\hat{\Theta}}\right) \tag{2.73}$$

Step 2. We derive the error dynamics z_2

$$\dot{z}_2 = \dot{v}_{m,2} - \dot{\alpha}_1 - \dot{\hat{\varrho}}_r - \hat{\varrho}\ddot{y}_r$$

$$= v_{m,3} - k_2 v_{m,1} - \frac{\partial\alpha_1}{\partial y}\left(b_m v_{m,2} + \xi_2 + \bar{\omega}^T\Theta + \epsilon_2\right) - \frac{\partial\alpha_1}{\partial y_r}\dot{y}_r$$

$$- \sum_{j=1}^{m+i-1}\frac{\partial\alpha_1}{\partial\lambda_j}\left(-k_j\lambda_1 + \lambda_{j+1}\right) - \frac{\partial\alpha_1}{\partial\xi}\left(A_0\xi + ky\right)$$

$$- \frac{\partial\alpha_1}{\partial\Xi}\left(A_0\Xi^T + \Phi(y)\right) - \frac{\partial\alpha_1}{\partial\hat{\Theta}}\dot{\hat{\Theta}} - \frac{\partial\alpha_1}{\partial\hat{\varrho}}\dot{\hat{\varrho}} - \hat{\varrho}\dot{y}_r - \hat{\varrho}\ddot{y}_r$$

$$= v_{m,3} - \hat{\varrho}\ddot{y}_r - \beta_2 - \frac{\partial\alpha_1}{\partial y}\left(\omega^T\tilde{\Theta} + \epsilon_2)\right) - \frac{\partial\alpha_1}{\partial\hat{\Theta}}\dot{\hat{\Theta}} \tag{2.74}$$

where

$$\beta_2 = \frac{\partial \alpha_1}{\partial y} \left(\xi_2 + \omega^T \hat{\Theta} \right) + k_2 v_{m,1} + \frac{\partial \alpha_1}{\partial y_r} \dot{y}_r + \left(\dot{y}_r + \frac{\partial \alpha_1}{\partial \hat{\varrho}} \right) \dot{\hat{\varrho}}$$

$$+ \sum_{j=1}^{m+i-1} \frac{\partial \alpha_1}{\partial \lambda_j} \left(-k_j \lambda_1 + \lambda_{j+1} \right) + \frac{\partial \alpha_1}{\partial \xi} \left(A_0 \xi + ky \right)$$

$$+ \frac{\partial \alpha_1}{\partial \Xi^T} \left(A_0 \Xi^T + \Phi(y) \right) \tag{2.75}$$

By considering $v_{m,3}$ as virtual control input and using $z_3 = v_{m,3} - \alpha_2 - \hat{\varrho} \ddot{y}_r$, we have

$$\dot{z}_2 = z_3 + \alpha_2 - \beta_2 - \frac{\partial \alpha_1}{\partial y} \left(\omega^T \tilde{\Theta} + \epsilon_2 \right) - \frac{\partial \alpha_1}{\partial \hat{\Theta}} \dot{\hat{\Theta}} \tag{2.76}$$

With the Lyapunov function

$$V_2 = V_1 + \frac{1}{2} z_2^2 + \frac{1}{2d_2} \epsilon^T P \epsilon, \tag{2.77}$$

we choose the second virtual control law α_2 and tuning function as

$$\alpha_2 = -\hat{b}_m z_1 - \left(c_2 + d_2 \left(\frac{\partial \alpha_1}{\partial y} \right)^2 \right) z_2 + \beta_2 + \frac{\partial \alpha_1}{\partial \hat{\Theta}} \Gamma \tau_2 \tag{2.78}$$

$$\tau_2 = \tau_1 - \frac{\partial \alpha_1}{\partial y} \omega z_2 \tag{2.79}$$

Then,

$$\dot{V}_2 = \dot{V}_1 + z_2 \dot{z}_2 - \frac{1}{2d_2} \epsilon^T \epsilon$$

$$\leq -c_1 z_1^2 + \hat{b}_m z_1 z_2 + z_2 \left(z_3 + \alpha_2 - \beta_2 - \frac{\partial \alpha_1}{\partial y} \left(\omega^T \tilde{\Theta} + \epsilon_2 \right) - \frac{\partial \alpha_1}{\partial \hat{\Theta}} \dot{\hat{\Theta}} \right)$$

$$- \frac{1}{2d_2} \epsilon^T \epsilon - \frac{1}{4d_1} \epsilon^T \epsilon + \tilde{\Theta}^T \left(\tau_1 - \Gamma^{-1} \dot{\hat{\Theta}} \right)$$

$$= -c_1 z_1^2 - c_2 z_2^2 + z_2 z_3 - d_2 \left(\frac{\partial \alpha_1}{\partial y} \right)^2 z_2^2 - \frac{\partial \alpha_1}{\partial y} \epsilon_2 z_2 - \frac{1}{4d_2} \epsilon^T \epsilon$$

$$- \frac{1}{4d_2} \epsilon^T \epsilon - \frac{1}{4d_1} \epsilon^T \epsilon + \tilde{\Theta}^T \left(\tau_1 - \frac{\partial \alpha_1}{\partial y} \omega z_2 - \Gamma^{-1} \dot{\hat{\Theta}} \right) + \frac{\partial \alpha_1}{\partial \hat{\Theta}} \left(\Gamma \tau_2 - \dot{\hat{\Theta}} \right)$$

$$\leq - \sum_{i=1}^{2} \left(c_i z_i^2 + \frac{1}{4d_i} \epsilon^T \epsilon \right) + z_2 z_3 + \tilde{\Theta}^T \left(\tau_2 - \Gamma^{-1} \dot{\hat{\Theta}} \right)$$

$$+ \frac{\partial \alpha_1}{\partial \hat{\Theta}} \left(\Gamma \tau_2 - \dot{\hat{\Theta}} \right) \tag{2.80}$$

Step $i(i = 3, \ldots, \rho)$. Choose virtual control laws

$$\alpha_i = -z_{i-1} - \left[c_i + d_i \left(\frac{\partial \alpha_{i-1}}{\partial y}\right)^2\right] z_i + \beta_i + \frac{\partial \alpha_{i-1}}{\partial \hat{\Theta}} \Gamma \tau_i$$

$$- \left(\sum_{k=2}^{i-1} z_k \frac{\partial \alpha_{k-1}}{\partial \hat{\Theta}}\right) \Gamma \frac{\partial \alpha_{i-1}}{\partial y} \omega, i = 3, \ldots, \rho \qquad (2.81)$$

where c_i are positive design parameters and

$$\tau_i = \tau_{i-1} - \frac{\partial \alpha_{i-1}}{\partial y} \omega z_i \qquad (2.82)$$

$$\beta_i = \frac{\partial \alpha_{i-1}}{\partial y} \left(\xi_2 + \omega^T \hat{\Theta}\right) + k_i v_{m,1} + \sum_{j=1}^{i-1} \frac{\partial \alpha_{i-1}}{\partial y_r^{(j-1)}} y_r^{(j)}$$

$$+ \left(y_r^{(i-1)} + \frac{\partial \alpha_{i-1}}{\partial \hat{\varrho}}\right) \dot{\hat{\varrho}} + \sum_{j=1}^{m+i-1} \frac{\partial \alpha_{i-1}}{\partial \lambda_j} \left(-k_j \lambda_1 + \lambda_{j+1}\right)$$

$$+ \frac{\partial \alpha_{i-1}}{\partial \xi} \left(A_0 \xi + ky\right) + \frac{\partial \alpha_{i-1}}{\partial \Xi^T} \left(A_0 \Xi^T + \Phi(y)\right) \qquad (2.83)$$

In the last step ρ, the adaptive controller and parameter update law are finally given by

$$u = \alpha_\rho - v_{m,\rho+1} + \hat{\varrho} y_r^{(\rho)} \qquad (2.84)$$

$$\dot{\hat{\Theta}} = \Gamma \tau_\rho \qquad (2.85)$$

We define the final Lyapunov function V_ρ as

$$V_\rho = \sum_{i=1}^{\rho} \frac{1}{2} z_i^2 + \frac{1}{2} \tilde{\Theta}^T \Gamma^{-1} \tilde{\Theta} + \frac{|b_m|}{2\gamma} \tilde{\varrho}^2 + \sum_{i=1}^{\rho} \frac{1}{2d_i} \epsilon^T P \epsilon \qquad (2.86)$$

Note that

$$\Gamma \tau_{i-1} - \dot{\hat{\Theta}} = \Gamma \tau_{i-1} - \Gamma \tau_i + \Gamma \tau_i - \dot{\hat{\Theta}} \qquad (2.87)$$

$$= \Gamma \frac{\partial \alpha_{i-1}}{\partial y} \omega z_i + \left(\Gamma \tau_i - \dot{\hat{\Theta}}\right) \qquad (2.88)$$

From (2.81)–(2.85), the derivative of the last Lyapunov function satisfies

$$\dot{V}_\rho = \sum_{i=1}^{\rho} z_i \dot{z}_i - \tilde{\Theta}^T \Gamma^{-1} \dot{\hat{\Theta}} - \frac{|b_m|}{\gamma} \tilde{\varrho} \dot{\hat{\varrho}} - \sum_{i=1}^{\rho} \frac{1}{2d_i} \epsilon^T \epsilon$$

$$\leq -\sum_{i=1}^{\rho} c_i z_i^2 - \sum_{i=1}^{\rho} \frac{1}{4d_i} \epsilon^T \epsilon - \tilde{\Theta}^T \Gamma^{-1} \left(\dot{\hat{\Theta}} - \Gamma \tau_\rho\right)$$

$$+ \left(\sum_{k=2}^{\rho} z_k \frac{\partial \alpha_{k-1}}{\partial \hat{\Theta}} \right) \left(\Gamma \tau_\rho - \dot{\hat{\Theta}} \right)$$

$$= - \sum_{i=1}^{\rho} c_i z_i^2 - \sum_{i=1}^{\rho} \frac{1}{4d_i} \epsilon^T \epsilon \qquad (2.89)$$

We have the following stability results based on the designed backstepping controller.

Theorem 2.2 *Consider the system (2.39) consisting of the parameter estimators given by (2.71) and (2.85), adaptive controllers designed using (2.84) with virtual control laws (2.64), (2.78), and (2.81), and the filters (2.42)–(2.44). Then all signals in the closed-loop system are globally uniformly bounded and asymptotic tracking is achieved, i.e., $\lim_{t\to\infty} [y(t) - y_r(t)] = 0$.*

Proof. Due to the piecewise continuity of $y_r(t), \ldots, y_r^{(\rho)}(t)$ and the smoothness of the control law, the parameter updating laws and the filters, the solution of the closed-loop adaptive system exists and is unique. From (2.89), it can be shown that V_ρ is uniformly bounded. Thus z_i, $\hat{\Theta}$, $\hat{\varrho}$, and ϵ are bounded. Since z_1 and y_r are bounded, y is also bounded. Then from (2.42) and (2.43) we conclude that ξ and Ξ are bounded as A_0 is Hurwitz. From (2.44) and Assumption 2, we have that $\lambda_1, \ldots, \lambda_{m+1}$ are bounded. From the coordinate change -(2.61), it gives

$$v_{m,i} = z_i + \hat{\varrho} y_r^{(i-1)} + \alpha_{i-1} \left(y, \xi, \Xi, \hat{\Theta}, \hat{\varrho}, \bar{\lambda}_{m+i-1}, \bar{y}_r^{(i-2)} \right), i = 2, 3, \ldots, \rho, \qquad (2.90)$$

where $\bar{\lambda}_k = [\lambda_1, \ldots, \lambda_k]^T$, $\bar{y}_r^{(k)} = \left[y_r, \ldots, y_r^{(k)} \right]^T$. For $i = 2$, from the boundedness of $\lambda_{m+1}, z_2, y, \xi, \Xi, \hat{\Theta}, \hat{\varrho}, y_r$, and $\dot{y}_r$, it proves that $v_{m,2}$ is bounded. From (2.45) it follows that λ_{m+2} is bounded. Following the same procedure recursively, we can show that λ is bounded. From (2.46) and the boundedness of $\xi, \Xi, \lambda, \epsilon$, we conclude that x is bounded.

To show the global uniform stability, the boundedness of $m = n - \rho$ dimension states ζ with zero dynamics should be guaranteed. Under a similar transformation as in [56], the states ζ associated with the zero dynamics can be shown to satisfy

$$\dot{\zeta} = A_b \zeta + b_b y + T\Phi(y)\theta \qquad (2.91)$$

where $\zeta = Tx, b_b \in \Re^m$, the eigenvalues of the $m \times m$ matrix A_b is given as follows

$$A_b = \begin{bmatrix} -b_{m-1}/b_m & & \\ & & I_{m-1} \\ \vdots & & \\ -b_0/b_m & 0 & \cdots & 0 \end{bmatrix} \qquad (2.92)$$

$$T = [(A_b)^\rho e_1, \ldots, A_b e_1, I_m] . \qquad (2.93)$$

With Assumption 2.3.4, we have that A_b is Hurwitz. Hence, there exists matrix P such that

$$PA_b + (A_b)^T P = -2I \tag{2.94}$$

Now we define a Lyapunov function for the zero dynamics of the system as $V_\zeta = \zeta^T P \zeta$. It can be shown that

$$\dot{V}_\zeta \le -\zeta^T \zeta + \left\| P \left(b_b y + T\Phi(y)^T \theta \right) \right\|^2 \tag{2.95}$$

Because all signals and functions in the second term of (2.95) are bounded, it can be shown that ζ is bounded.

Thus all signals in the closed-loop are globally uniformly bounded. By applying the LaSalle-Yoshizawa theorem to (2.89), it further follows that $z(t) \to 0$ as $t \to \infty$, which implies that $\lim_{t\to\infty}[y(t) - y_r(t)] = 0$.

2.4 DISTRIBUTED ADAPTIVE CONSENSUS CONTROL

In this section, we first discuss the leader-following consensus control problem of first-order nonlinear multi-agent systems with unknown parameters. Under undirected graph condition, a distributed adaptive consensus tracking control scheme is presented. Along the same design procedure, we then try to extend the results to the directed graph case. Finally, similar to the leader-following consensus case, the leaderless consensus control problem for uncertain first-order nonlinear multi-agent systems will be discussed under both undirected and directed graph conditions.

2.4.1 LEADER-FOLLOWING CONSENSUS CONTROL

Consider a group of N first-order nonlinear multi-agent systems, where the dynamics of the ith agent is described as follows.

$$\dot{x}_i = u_i + \varphi_i(x_i)\theta_i, \tag{2.96}$$
$$y_i = x_i, i = 1, \ldots, N, \tag{2.97}$$

where $x_i \in \Re$, $u_i \in \Re$, and $y_i \in \Re$ are the state, control input and output of agent i, respectively, $\theta_i \in \Re$ is an unknown parameter, $\varphi_i(x_i) \in \Re$ is a known nonlinear function. For notational convenience, $\varphi_i(x_i)$ is simplified as φ_i in this section. The desired trajectory is generated by a function $y_r(t) = 1$ and at least one agent has access to the desired trajectory. We use $\mu_i = 1$ to denote the case that the desired trajectory $y_r(t)$ is available for agent i, otherwise $\mu_i = 0$.

The control objective is to design a distributed adaptive controller for each agent such that its output can track the desired trajectory $y_r(t)$ asymptotically, though only a portion of agents have access to the desired trajectory $y_r(t)$. That is, $\lim_{t\to\infty}[y_i(t) - y_r(t)] = 0$.

To design the distributed adaptive controller, the following error variables are first introduced

$$z_i = \sum_{j=1}^{N} a_{ij}(y_i - y_j) + \mu_i(y_i - y_r), \qquad (2.98)$$

$$\delta_i = y_i - y_r, \qquad (2.99)$$

where z_i is the local leader-following consensus error and δ_i is the actual tracking error. Let $z = [z_1, \ldots, z_N]^T$, then $z = (\mathcal{L} + \mathcal{B})\delta$ where $\mathcal{L}$ is the Laplacian matrix, $\mathcal{B} = \text{diag}\{\mu_1, \ldots, \mu_N\}$ and $\delta = [\delta_1, \ldots, \delta_N]^T$.

The time-derivative of z is computed as

$$\dot{z} = (\mathcal{L} + \mathcal{B})\dot{\delta} = (\mathcal{L} + \mathcal{B}) \begin{bmatrix} u_1 + \varphi_1\theta_1 \\ \vdots \\ u_N + \varphi_N\theta_N \end{bmatrix}. \qquad (2.100)$$

Now, we separately design the distributed adaptive consensus tracking controller for undirected and directed graph cases.

◇ *Undirected Graph Case*: Assume that the undirected graph is connected. From Lemma 2.1, the matrix $\mathcal{L}+\mathcal{B}$ is symmetric positive definite with $\mathcal{L}+\mathcal{B} = \mathcal{L}^T+\mathcal{B}$. To design a suitable consensus tacking controller for each agent, the Lyapunov function for the entire closed-loop system is defined as

$$V = \frac{1}{2}\delta^T(\mathcal{L} + \mathcal{B})\delta + \frac{1}{2}\sum_{i=1}^{N} \tilde{\theta}_i^2, \qquad (2.101)$$

where $\tilde{\theta}_i = \theta_i - \hat{\theta}_i$ and $\hat{\theta}_i$ are the estimation error and estimate of the unknown parameter θ_i, respectively.

The derivative of (2.101) can be computed as

$$\dot{V} = z^T\dot{\delta} + \sum_{i=1}^{N} \tilde{\theta}_i(-\dot{\hat{\theta}}_i) = z^T \begin{bmatrix} u_1 + \varphi_1\theta_1 \\ \vdots \\ u_N + \varphi_N\theta_N \end{bmatrix} + \sum_{i=1}^{N} \tilde{\theta}_i(-\dot{\hat{\theta}}_i). \qquad (2.102)$$

The distributed controller and parameter update law for agent i are chosen as

$$u_i = -z_i - \varphi_i\hat{\theta}_i, \qquad (2.103)$$

$$\dot{\hat{\theta}}_i = \varphi_i z_i, \qquad (2.104)$$

Substituting (2.103)–(2.104) into (2.102), the derivative of V can be further derived as

$$\dot{V} = -\sum_{i=1}^{N} z_i^2. \qquad (2.105)$$

It can be concluded from (2.105) that z_i and $\hat{\theta}_i$ are bounded and $z_i \in \mathcal{L}_2$. By applying the LaSalle-Yoshizawa theorem, we have $\lim_{t\to\infty} z_i(t) = 0$. Since $z = (\mathcal{L}+\mathcal{B})\delta$ and $\mathcal{L} + \mathcal{B}$ is a positive definite matrix, we can further get that $\lim_{t\to\infty} \delta_i(t) = 0$. This means that with the designed distributed controller (2.103) and parameter update law (2.104), the output of each agent can track the desired trajectory $y_r(t) = 1$ asymptotically.

◇ *Directed Graph Case*: Assume that the directed graph is weakly connected and balanced graph. For such graph, the matrix $\mathcal{L}+\mathcal{B}$ may be asymmetric, that is, $\mathcal{L}+\mathcal{B} \neq \mathcal{L}^T +\mathcal{B}$. However, Lemma 2.3 shows that $\mathcal{L}+\mathcal{L}^T +2\mathcal{B}$ is symmetric positive definite. Now, we try to use these properties to design the distributed adaptive consensus tracking controller.

Since $\mathcal{L} + \mathcal{B}$ may not be positive definite, the Lyapunov function V in (2.101) is modified as

$$V = \frac{1}{2}\sum_{i=1}^{N} z_i^2 + \frac{1}{2}\sum_{i=1}^{N} \tilde{\theta}_i^2. \tag{2.106}$$

The derivative of V can be computed as

$$\dot{V} = z^T(\mathcal{L}+\mathcal{B}) \begin{bmatrix} u_1 + \varphi_1\theta_1 \\ \vdots \\ u_N + \varphi_N\theta_N \end{bmatrix} + \sum_{i=1}^{N} \tilde{\theta}_i(-\dot{\hat{\theta}}_i). \tag{2.107}$$

If the distributed adaptive controller is also chosen as $u_i = -z_i - \varphi_i\hat{\theta}_i$, then

$$\dot{V} = -\frac{1}{2}z^T(\mathcal{L}+\mathcal{L}^T+2\mathcal{B})z + z^T(\mathcal{L}+\mathcal{L}^T+2\mathcal{B}) \begin{bmatrix} \varphi_1\tilde{\theta}_1 \\ \vdots \\ \varphi_N\tilde{\theta}_N \end{bmatrix} + \sum_{i=1}^{N} \tilde{\theta}_i(-\dot{\hat{\theta}}_i). \tag{2.108}$$

From (2.108), we can observe that the coupling term associated with local consensus error and parameter estimation error exists. This brings the main difficulty on designing a distributed parameter update law. How to solve this issue is interesting. Some valid distributed consensus control schemes will be proposed in the subsequent chapters in this monograph.

2.4.2 LEADERLESS CONSENSUS CONTROL

Similar to the leader-following consensus case, we investigate the leaderless consensus control problem under undirected and directed graphs and explore what interesting things will arise.

Consider the same nonlinear multi-agent systems as in (2.96). The control objective is to design a distributed adaptive leaderless consensus controller for each agent such that the states of all the agents can reach a common value, that is, $\lim_{t\to\infty}[x_i(t) - x_j(t)] = 0$, $i, j \in \mathcal{V}$.

Let $x = [x_1, \ldots, x_N]^T$ and define a local leaderless consensus error as $z_i = \sum_{j=1}^{N}(x_i - x_j)$. We first consider the undirected graph case.

◇ *Undirected Graph Case*: Assume that the undirected graph is connected, then the matrix $\mathcal{L}$ is symmetric semi-positive definite with $\mathcal{L} = \mathcal{L}^T$.

Let $\tilde{\theta}_i = \theta_i - \hat{\theta}_i$ and $\hat{\theta}_i$ be, respectively, the estimation error and estimate of the unknown parameter θ_i. The Lyapunov function for the entire closed-loop system is defined as

$$V = \frac{1}{2}x^T \mathcal{L} x + \frac{1}{2}\sum_{i=1}^{N} \tilde{\theta}_i^2, \qquad (2.109)$$

From (2.96), the derivative of (2.109) can be computed as

$$\dot{V} = x^T \mathcal{L}\dot{x} + \sum_{i=1}^{N} \tilde{\theta}_i(-\dot{\hat{\theta}}_i) = x^T \mathcal{L} \begin{bmatrix} u_1 + \varphi_1\theta_1 \\ \vdots \\ u_N + \varphi_N\theta_N \end{bmatrix} + \sum_{i=1}^{N} \tilde{\theta}_i(-\dot{\hat{\theta}}_i). \qquad (2.110)$$

If the distributed controller and parameter update law are chosen as

$$u_i = -x_i - \varphi_i\hat{\theta}_i, \qquad (2.111)$$

$$\dot{\hat{\theta}}_i = \varphi_i z_i, \qquad (2.112)$$

then the derivative of (2.110) can be further derived as

$$\dot{V} = -\sum_{i=1}^{N}\sum_{j=1}^{N}(x_i - x_j)^2. \qquad (2.113)$$

By applying the LaSalle theorem, it can be concluded from (2.113) that $\lim_{t\to\infty}[x_i(t) - x_j(t)] = 0, i, j \in \mathcal{V}$, which means that the asymptotically leaderless consensus is achieved.

Hence, from the above design procedure, we know that the distributed adaptive leaderless consensus control scheme can be developed under undirected graph condition. However, it is difficulty to analyze the final consensus value based only on the equation (2.113), due to the complicated nonlinear system dynamics.

Now, we turn to the directed graph case and explore what obstacles will arise.

◇ *Directed Graph Case*: Assume that the directed graph has a directed spanning tree. Under such condition, the Laplacian matrix $\mathcal{L}$ may not be semi-positive definite. Given this reason, the Lyapunov function (2.109) is redefined as

$$V = \frac{1}{2}\sum_{i=1}^{N} z_i^2 + \frac{1}{2}\sum_{i=1}^{N} \tilde{\theta}_i^2, \qquad (2.114)$$

where z_i is the local leaderless consensus error, $\tilde{\theta}_i = \theta_i - \hat{\theta}_i$ and $\hat{\theta}_i$ are, respectively, the estimation error and estimate of the unknown parameter θ_i. Then, the derivative of (2.114) can be computed as

$$
\dot{V} = \sum_{i=1}^{N} z_i \dot{z}_i + \sum_{i=1}^{N} \tilde{\theta}_i(-\dot{\hat{\theta}}_i)
$$

$$
= \sum_{i=1}^{N} \sum_{j=1}^{N} z_i(u_i + \varphi_i \theta_i - u_j - \varphi_j \theta_j) + \sum_{i=1}^{N} \tilde{\theta}_i(-\dot{\hat{\theta}}_i). \tag{2.115}
$$

If the distributed adaptive controller is also chosen as $u_i = -x_i - \varphi_i \hat{\theta}_i$, then

$$
\dot{V} = -\sum_{i=1}^{N} z_i^2 + \sum_{i=1}^{N} \sum_{j=1}^{N} a_{ij} z_i(\varphi_i \tilde{\theta}_i - \varphi_j \tilde{\theta}_j) + \sum_{i=1}^{N} \tilde{\theta}_i(-\dot{\hat{\theta}}_i). \tag{2.116}
$$

Since the directed graph has a directed spanning tree and is nonsymmetric, $\sum_{i=1}^{N} a_{ij} \neq \sum_{j=1}^{N} a_{ij}$. It can be observed from (2.116) that the coupling terms associated with local consensus error and parameter estimation error also exist, which makes the design of distributed parameter estimators difficult.

2.4.3 DISCUSSION

In this section, we investigate the leader-following and leaderless consensus control problems for first-order nonlinear multi-agent systems with unknown parameters. For the both cases, distributed adaptive consensus control laws with suitable parameter update laws can be elaborately designed under undirected graph condition. However, how to design a distributed consensus control scheme to achieve leader-following or leaderless consensus under directed graph condition is nontrivial since the Laplacian matrix $\mathcal{L}$ is nonsymmetric. From the detail derivation processes above, we can see that the main difficulty lies in that the coupling term associated with local consensus error and parameter estimation error unavoidably exists if the Lyapunov function is chosen based on the local consensus error z_i. This hinders the design of fully distributed parameter update laws. To overcome this difficulty, some novel methods such as redefining a suitable Lyapunov function or introducing auxiliary system for each agent will be presented in the subsequent chapters of this monograph, by considering a class of more general uncertain high-order nonlinear multi-agent systems.

2.5 NOTES

Some preliminaries on notations, algebraic graph theory, adaptive backstepping control technique, and distributed adaptive consensus control are introduced in this chapter. It should be mentioned that the content of distributed adaptive consensus control is originated from our review paper [132].

Section I

Consensus Control under Directed Graph

3 Distributed Adaptive State Feedback Control with Linearly Parametric Reference

In Chapter 1, we discussed the importance and significance of the investigation of consensus control for uncertain nonlinear multi-agent systems. In Chapter 2, the basic design procedure of consensus control laws for first-order nonlinear multi-agent systems is introduced and the main difficulties on designing consensus control laws under directed graph condition are pointed out. More results on this topic can also refer to some representative works, see [16, 17, 41, 42, 51, 72, 77, 90, 121, 148, 155, 161, 167] for instance. However, it should be noted that the results on distributed adaptive consensus control of more general multiple high-order nonlinear systems are still limited. In [139], output consensus tracking problem for nonlinear agents in the presence of mismatched unknown parameters is investigated. By designing an estimator whose dynamics is governed by a chain of n integrators for the desired trajectory in each agent, bounded output consensus tracking for the overall system can be achieved. However, it is not easy to check whether the derived sufficient condition in the form of Linear Matrix Inequality (LMI) is satisfied by choosing the design parameters properly. Moreover, transmissions of online parameter estimates among the neighbors are required, which may increase communication burden and also cause some other potential problems such as those related to network security.

In this chapter, we investigate the output consensus tracking control problem of a class of systems consisting of multiple nonlinear agents with intrinsic mismatched unknown parameters. The agents are allowed to have non-identical dynamics, with a similar structure and the same yet arbitrary system order. The communication status among the agents can be represented by a directed graph. Inspired by [5, 6, 155], the desired trajectory is assumed to be linearly parameterized. Different from the traditional centralized tracking control problem, only a subset of the agents can obtain the desired trajectory information directly. A distributed adaptive control approach based on backstepping technique is proposed. By introducing the estimates to account for the parametric uncertainties of the desired trajectory and its neighbors' dynamics into the local controller of each agent, information exchanges of online parameter estimates and local synchronization errors among linked agents can be avoided. It is proved that the boundedness of all closed-loop signals and the asymptotically consensus tracking for all the agents' outputs are ensured.

DOI: 10.1201/9781003394372-3

3.1 PROBLEM FORMULATION

We consider a group of N nonlinear agents which can be modeled as follows.

$$\dot{x}_{i,q} = x_{i,q+1} + \varphi_{i,q}(x_{i,1}, \ldots, x_{i,q})^T \theta_i, \quad q = 1, \ldots, n-1$$
$$\dot{x}_{i,n} = b_i \beta_i(x_i) u_i + \varphi_{i,n}(x_i)^T \theta_i,$$
$$y_i = x_{i,1}, \quad \text{for } i = 1, 2, \ldots, N, \tag{3.1}$$

where $x_i = [x_{i,1}, \ldots, x_{i,n}]^T \in \Re^n$, $u_i \in \Re$ and $y_i \in \Re$ are the states, control input, and output of the ith agent, respectively. $\theta_i \in \Re^{p_i}$ is a vector of unknown constants and the high frequency gain $b_i \in \Re$ is an unknown non-zero constant. $\varphi_{i,j} : \Re^j \to \Re^{p_i}$ for $j = 1, \ldots, n$ and $\beta_i : \Re^n \to \Re^1$ are known smooth nonlinear functions.

The desired trajectory for the outputs of the overall system can be expressed by a linear combination of q_r basis functions, that is

$$y_r(t) = \sum_{l=1}^{q_r} f_{r,k}(t) w_{r,l} + c_r = f_r(t)^T w_r + c_r, \tag{3.2}$$

where $f_r(t) = [f_{r,1}(t), f_{r,2}(t), \ldots, f_{r,q_r}(t)]^T \in \Re^{q_r}$ is the vector of basis functions which is available to all the N agents. However, $w_r = [w_{r,1}, w_{r,2}, \ldots, w_{r,q_r}]^T \in \Re^{q_r}$ and $c_r \in \Re$ are constant parameters which are known only to part of N agents.

Remark 3.1 *It is worth mentioning that the trajectory given in (3.2) is a commonly employed expression which has appeared in many relevant literature such as [5, 6, 155]. As we know, a function can be represented or approximated as a linear combination of a set of prescribed basis functions in a function space. For example, if a desired trajectory $y_r(t)$ is periodic with period T, then $y_r(t)$ can be written as $y_r(t) = a_0 + \sum_{k=1}^{\infty} \left(a_k \cos \frac{2\pi kt}{T} + b_k \sin \frac{2\pi kt}{T}\right)$. This is known as trigonometric form of the Fourier series, in which a_0, a_k, and b_k are constants called Fourier coefficients [54]. If there are only finite dominant frequency components in $y_r(t)$, in other words, the contributions of $a_k \cos \frac{2\pi kt}{T}$ and $b_k \sin \frac{2\pi kt}{T}$ are negligible for $k > K$, then $y_r(t)$ can be approximated well by $\hat{y}_r(t) = a_0 + \sum_{k=1}^{K} \left(a_k \cos \frac{2\pi kt}{T} + b_k \sin \frac{2\pi kt}{T}\right)$ which has a similar form as (3.2).*

Remark 3.2 *In contrast to the traditional centralized trajectory tracking control problem, not all agents in the group can obtain exact knowledge of the trajectory $y_r(t)$ directly. However, the consensus tracking to $y_r(t)$ for all the N agents' outputs can still be expected if the agents are able to share information with some others in their neighboring areas via communication networks.*

We now use $\mu_i = 1$ to indicate the case that $y_r(t)$ is accessible directly to agent i; otherwise, μ_i is set as $\mu_i = 0$. Based on this, the *control objective* is to design distributed adaptive controllers u_i for each agent by utilizing only locally available information obtained from the intrinsic agent and its neighbors such that:

- All the signals in the closed-loop system are globally uniformly bounded;
- The outputs of the overall system can still track the desired trajectory $y_r(t)$ asymptotically, i.e., $\lim_{t\to\infty}[y_i(t)-y_r(t)] = 0$, $\forall i \in \{1, 2, \ldots, N\}$, though $\mu_i = 1$ only for some agents.

To achieve the objective, the following assumptions are imposed.

Assumption 3.1.1 *The first nth-order derivatives of $f_r(t)$ are bounded, piecewise continuous, and known to all agents in the group.*

Assumption 3.1.2 *The sign of b_i is available in constructing u_i for agent i and $\beta_i(x_i) \neq 0$.*

Assumption 3.1.3 *The directed graph $\mathcal{G}$ contains a spanning tree and the root node i has direct access to y_r, i.e., $\mu_i = 1$.*

Remark 3.3 *Observing (3.1), the system model considered in this chapter is similar to that in [139]. Such a model is more general than those in most of the currently available results on distributed consensus control including [2, 5, 6, 40, 87, 105, 114, 116, 157] by combining the following features: i) the agents are nonlinear and allowed to have non-identical dynamics; ii) intrinsic mismatched unknown parameters are involved. Moreover, (3.1) is in the parametric strict feedback form, which can be commonly encountered in many nonlinear control problems. In [56], the conditions that a class of general nonlinear systems $\dot{\chi} = f_0(\chi)+\sum_{l=1}^{p} \theta_l f_l(x)+ g(\chi)u, y = h(\chi)$ are transformable into such form have been provided. It should be noted that Assumptions 3.1.1 and 3.1.2 are also required for standard backstepping based centralized adaptive tracking control.*

3.2 AN ILLUSTRATIVE EXAMPLE

In Section 2.4, we pointed out that designing a distributed consensus control scheme to achieve leader-following consensus under directed graph condition is nontrivial if the Lyapunov function is defined using the local consensus error. To address this issue, we propose that the Lyapunov function can be chosen based on a well-defined local tracking error in this chapter. Now, we consider a special case of the multi-agent systems in (3.1) (i.e., $n = 1$) as an example to introduce the design idea. We rewrite the system dynamics of the N first-order nonlinear multi-agent systems as follows

$$\dot{x}_{i,1} = u_i + \varphi_{i,1}(x_{i,1})^T \theta_i,$$
$$y_i = x_{i,1}, \quad \text{for } i = 1, 2, \ldots, N. \tag{3.3}$$

⋄ *Distributed Adaptive Controller Design*: In what follows, we set out to design the distributed adaptive leader-following consensus controller for each agent. Suppose that Assumptions 3.1.1–3.1.3 hold.

For agent i with $\mu_i = 0$, we introduce $\hat{\bar{w}}_{ri} = [\hat{w}_{ri}^T, \hat{c}_{ri}]^T$ to estimate the unknown parameters w_r and c_r. Then the following error variables are defined

$$e_{i,1} = y_i - \mu_i y_r - (1 - \mu_i)\bar{f}_r^T \hat{\bar{w}}_{ri}, \tag{3.4}$$

$$z_i = \sum_{j=1}^{N} a_{ij}(y_i - y_j) + \mu_i(y_i - y_r), \tag{3.5}$$

where $\bar{f}_r = [f_r^T, 1]^T$. The actual tracking errors between each agents' outputs and y_r are defined as $\delta_i = y_i - y_r$, for $i = 1, \ldots, N$. Clearly, the control objective is to ensure that $\lim_{t \to \infty} \delta_i(t) = 0$ for all agents in the group. Equation (3.5) is a standard definition of the local consensus error. By defining $z = [z_1, \ldots, z_N]^T$, we have

$$z = (\mathcal{L} + \mathcal{B})\,\delta, \tag{3.6}$$

where $\delta = [\delta_1, \ldots, \delta_N]^T$.

From (3.2) and (3.4),

$$e_{i,1} = y_i - y_r + (1 - \mu_i)\left(y_r - \bar{f}_r^T \hat{\bar{w}}_{ri}\right)$$
$$= \delta_i + (1 - \mu_i)\bar{f}_r^T \tilde{\bar{w}}_{ri}, \tag{3.7}$$

where $\tilde{\bar{w}}_{ri}$ denotes the estimation error for agents with $\mu_i = 0$ such that $\tilde{\bar{w}}_{ri} = [w_r^T, c_r]^T - \hat{\bar{w}}_{ri}$.

From (3.3) and (3.4), the derivative of $e_{i,1}$ is computed as

$$\dot{e}_{i,1} = u_i + \varphi_{i,1}^T \theta_i - \mu_i \dot{f}_r^T w_r - (1 - \mu_i)\left(\dot{f}_r^T \hat{w}_{ri} + \bar{f}_r^T \dot{\hat{\bar{w}}}_{ri}\right). \tag{3.8}$$

We design u_i as

$$u_i = -k_1 P_i z_i - \varphi_{i,1}^T \hat{\theta}_i + \mu_i \dot{f}_r^T w_r + (1 - \mu_i)\left(\dot{f}_r^T \hat{w}_{ri} + \bar{f}_r^T \dot{\hat{\bar{w}}}_{ri}\right), \tag{3.9}$$

where k_1 is a positive constant, P_i is defined in Lemma 2.4, and $\hat{\theta}_i$ is the parameter estimate of θ_i.

The parameter update laws for $\hat{\theta}_i$ and $\hat{\bar{w}}_{ri}$ are chosen as

$$\dot{\hat{\theta}}_i = \Gamma_i \varphi_{i,1} e_{i,1}, \tag{3.10}$$

$$\dot{\hat{w}}_{ri} = -\Gamma_{ri}\bar{f}_r z_i, \tag{3.11}$$

where Γ_i and Γ_{ri} are positive definite matrices with appropriate dimensions.

$\diamond$ *Stability Analysis*: Now, we start to analyze the stability of the entire closed-loop system. The Lyapunov function for the entire closed-loop system is defined as

$$V_1 = \frac{1}{2}e_1^T e_1 + \frac{1}{2}\sum_{i=1}^{N}\tilde{\theta}_i^T\Gamma_i^{-1}\tilde{\theta}_i + \frac{k_1}{2}\sum_{i=1}^{N}(1-\mu_i)P_i\tilde{w}_{ri}^T\Gamma_{ri}^{-1}\tilde{w}_{ri}, \tag{3.12}$$

where $e_1 = [e_{1,1},\ldots,e_{N,1}]^T$, $\tilde{\theta}_i = \theta_i - \hat{\theta}_i$. Note that the Lyapunov function (3.12) is defined based on the well-defined local tracking error $e_{i,1}$, instead of the local consensus error z_i.

From (3.6)–(3.9), the derivative of V_1 is computed as

$$\dot{V}_1 = -k_1\delta^T P\left(\mathcal{L}+\mathcal{B}\right)\delta + \sum_{i=1}^{N}\tilde{\theta}_i^T\Gamma_i^{-1}\left(\Gamma_i\varphi_{i,1}e_{i,1} - \dot{\hat{\theta}}_i\right)$$

$$+ k_1\sum_{i=1}^{N}(1-\mu_i)p_i\tilde{w}_{ri}^T\Gamma_{ri}^{-1}\left(-\Gamma_{ri}\bar{f}_r z_i - \dot{\hat{w}}_{ri}\right). \tag{3.13}$$

Substituting (3.10) and (3.11) into (3.13), we have

$$\dot{V}_1 = -\frac{k_1}{2}\delta^T\left[P(\mathcal{L}+\mathcal{B}) + (\mathcal{L}+\mathcal{B})^T P\right]\delta$$

$$= -\frac{k_1}{2}\delta^T Q\delta, \tag{3.14}$$

where Q is a definite positive matrix defined in Lemma 2.4.

From (3.14), it follows that $e_{i,1}, \hat{\theta}_i$, and $\hat{w}_{ri}$ are bounded for all agent i. From (3.4) and the boundedness of f_r given in Assumption 3.1.1, we obtain that y_i, i.e., $x_{i,1}$ for $i = 1,\ldots,N$ are bounded. From (3.6) and $\delta = y - \underline{y}_r$, z is also bounded. From (3.9) and the smoothness of φ_i, u_i for $i = 1,\ldots,N$ are bounded. Thus the boundedness of all the signals in the closed-loop system is guaranteed. By applying the LaSalle-Yoshizawa theorem, it further follows from (3.14) that $\lim_{t\to\infty}\delta_i(t) = 0$ for $i = 1,\ldots,N$. Therefore, the leader-following consensus of the N first-order nonlinear multi-agent systems is achieved, i.e., $\lim_{t\to\infty}[y_i(t) - y_r(t)] = 0$ for $i = 1,\ldots,N$.

This example shows that the distributed adaptive leader-following consensus control problem for the first-order nonlinear multi-agent systems with directed communication topology can be solved, by resorting to the method of defining a new local tracking error $e_{i,1}$ in (3.7) and choosing a Lyapunov function associated with $e_{i,1}$ in (3.12).

3.3 DESIGN OF DISTRIBUTED ADAPTIVE CONTROLLERS

In this section, we further extend the results in Section 3.2 to the nth-order nonlinear multi-agent systems based on backstepping technique [56]. Although our design and analysis follow a step-by-step procedure under the general framework of backstepping, the details involved vary a lot in solving our problems.

Step 1. The design procedure in the first step is almost the same as that in Section 3.2. For brevity, only the key steps are provided. Except for the error variables $e_{i,1}$ and z_i defined in (3.4) and (3.5), we further introduce an error variable $e_{i,2}$ as

$$e_{i,2} = x_{i,2} - \alpha_{i,1}, \tag{3.15}$$

where $\alpha_{i,1}$ is a virtual control to be chosen.
The virtual control $\alpha_{i,1}$ is designed as

$$\alpha_{i,1} = -k_1 P_i z_i - \varphi_{i,1}^T \hat{\theta}_i + \mu_i \dot{f}_r^T w_r + (1 - \mu_i)\left(\dot{f}_r^T \hat{w}_{ri} + \bar{f}_r^T \dot{\hat{w}}_{ri}\right), \tag{3.16}$$

where k_1 is a positive constant, P_i is defined in Lemma 2.4, $\hat{\theta}_i$ is the parameter estimate of θ_i, and $\hat{w}_{ri} = [\hat{w}_{ri}^T, \hat{c}_{ri}]^T$ is the estimate of the unknown parameters w_r and c_r for the agent i with $\mu_i = 0$.
The parameter update law for $\hat{w}_{ri}$ if $\mu_i = 0$ is chosen as

$$\dot{\hat{w}}_{ri} = -\Gamma_{ri}\bar{f}_r z_i. \tag{3.17}$$

where Γ_{ri} is a positive definite matrix.
The Lyapunov function in this step is defined as

$$V_1 = \frac{1}{2}e_1^T e_1 + \frac{1}{2}\sum_{i=1}^N \tilde{\theta}_i^T \Gamma_i^{-1}\tilde{\theta}_i + \frac{k_1}{2}\sum_{i=1}^N (1 - \mu_i)P_i \tilde{w}_{ri}^T \Gamma_{ri}^{-1}\tilde{w}_{ri}, \tag{3.18}$$

where $e_1 = [e_{1,1}, \ldots, e_{N,1}]^T$, $\tilde{\theta}_i = \theta_i - \hat{\theta}_i$. Γ_i is positive definite matrix with appropriate dimension.
Define $\delta_i = y_i - y_r$, then $\delta = [\delta_1, \ldots, \delta_N]^T$ and $z = (\mathcal{L} + \mathcal{B})\delta$. Let $\tau_{i,1} = \varphi_{i,1}e_{i,1}$. From (3.15)–(3.17), the derivative of V_1 can be derived as

$$\dot{V}_1 = -\frac{k_1}{2}\delta^T \left[P(\mathcal{L} + \mathcal{B}) + (\mathcal{L} + \mathcal{B})^T P\right]\delta + \sum_{i=1}^N \left[e_{i,1}e_{i,2} + \tilde{\theta}_i^T \Gamma_i^{-1}\left(\Gamma_i \tau_{i,1} - \dot{\hat{\theta}}_i\right)\right]$$

$$= -\frac{k_1}{2}\delta^T Q\delta + \sum_{i=1}^N \left[e_{i,1}e_{i,2} + \tilde{\theta}_i^T \Gamma_i^{-1}\left(\Gamma_i \tau_{i,1} - \dot{\hat{\theta}}_i\right)\right]. \tag{3.19}$$

where Q is a positive definite matrix defined in Lemma 2.4.

Step 2. We now clarify the arguments of $\alpha_{i,1}$. By examining (3.16) along with (3.17), it can be seen that $\alpha_{i,1}$ is a function of y_i, $\hat{\theta}_i$, f_r, $\dot{f}_r$, y_j (if $a_{ij} = 1$) and $\hat{w}_{ri}$ (if $\mu_i = 0$). Introduce a new error variable as

$$e_{i,3} = x_{i,3} - \alpha_{i,2}. \tag{3.20}$$

From (3.1), (3.15), (3.16), and (3.20), the derivative of $e_{i,2}$ is computed as

$$
\dot{e}_{i,2} = e_{i,3} + \alpha_{i,2} + \left[\left(\varphi_{i,2} - \frac{\partial \alpha_{i,1}}{\partial x_{i,1}} \varphi_{i,1} \right)^T \theta_i - \sum_{j=1}^{N} a_{ij} \frac{\partial \alpha_{i,1}}{\partial x_{j,1}} \left(x_{j,2} + \varphi_{j,1}^T \theta_j \right) \right]
$$
$$
- \left[\frac{\partial \alpha_{i,1}}{\partial x_{i,1}} x_{i,2} + \frac{\partial \alpha_{i,1}}{\partial \hat{\theta}_i} \dot{\hat{\theta}}_i + \frac{\partial \alpha_{i,1}}{\partial f_r} \dot{f}_r + \frac{\partial \alpha_{i,1}}{\partial \dot{f}_r} \ddot{f}_r + (1 - \mu_i) \frac{\partial \alpha_{i,1}}{\partial \hat{\tilde{w}}_{ri}} \dot{\hat{\tilde{w}}}_{ri} \right],
$$
$$(3.21)$$

where $\alpha_{i,2}$ is chosen as

$$
\alpha_{i,2} = - e_{i,1} - k_{i,2} e_{i,2} + \frac{\partial \alpha_{i,1}}{\partial x_{i,1}} x_{i,2} + \frac{\partial \alpha_{i,1}}{\partial \hat{\theta}_i} \Gamma_i \tau_{i,2} - \left(\varphi_{i,2} - \frac{\partial \alpha_{i,1}}{\partial x_{i,1}} \varphi_{i,1} \right)^T \hat{\theta}_i
$$
$$
+ \sum_{j=1}^{N} a_{ij} \frac{\partial \alpha_{i,1}}{\partial x_{j,1}} \left(x_{j,2} + \varphi_{j,1}^T \hat{\theta}_{ij} \right) + \frac{\partial \alpha_{i,1}}{\partial f_r} \dot{f}_r
$$
$$
+ \frac{\partial \alpha_{i,1}}{\partial \dot{f}_r} \ddot{f}_r + (1 - \mu_i) \frac{\partial \alpha_{i,1}}{\partial \hat{\tilde{w}}_{ri}} \dot{\hat{\tilde{w}}}_{ri}.
$$
$$(3.22)$$

with $k_{i,2}$ a positive constant. $\tau_{i,2}$ is a tuning function defined as follows for generating $\dot{\hat{\theta}}_i$ that

$$
\tau_{i,2} = \tau_{i,1} + \left(\varphi_{i,2} - \frac{\partial \alpha_{i,1}}{\partial x_{i,1}} \varphi_{i,1} \right) e_{i,2}.
$$
$$(3.23)$$

$\hat{\theta}_{ij}$ is an estimator introduced in agent i to account for the unknown parameter vector contained in its neighbors' dynamics (i.e., θ_j if $a_{ij} = 1$).
From (3.15) and (3.20)–(3.23), the derivative of $e_{i,2}$ is computed as

$$
\dot{e}_{i,2} = - e_{i,1} - k_{i,2} e_{i,2} + e_{i,3} + \left(\varphi_{i,2} - \frac{\partial \alpha_{i,1}}{\partial x_{i,1}} \varphi_{i,1} \right)^T \tilde{\theta}_i
$$
$$
+ \frac{\partial \alpha_{i,1}}{\partial \hat{\theta}_i} \left(\Gamma_i \tau_{i,2} - \dot{\hat{\theta}}_i \right) - \sum_{j=1}^{N} a_{ij} \frac{\partial \alpha_{i,1}}{\partial x_{j,1}} \varphi_{j,1}^T \tilde{\theta}_{ij}.
$$
$$(3.24)$$

Remark 3.4 *The first term appearing in the second line of (3.22) implies that in constructing the local controller of the ith agent, $\varphi_{j,1}$ involving the structural knowledge of its neighbors' intrinsic dynamics is required if $a_{ij} = 1$. Similar requirement can also be found in [139]. However, in contrast to [139], the extra information exchange of local parameter estimates can be avoided in this chapter by introducing $\hat{\theta}_{ij}$ in the ith agent to estimate the unknown parameters contained in its neighbors' dynamics (i.e., θ_j).*

Define a Lyapunov function candidate V_2 at this step as

$$V_2 = V_1 + \frac{1}{2}\sum_{i=1}^{N} e_{i,2}^2 + \frac{1}{2}\sum_{i=1}^{N}\sum_{j=1}^{N} a_{ij}\tilde{\theta}_{ij}^T\Gamma_{ij}^{-1}\tilde{\theta}_{ij}, \tag{3.25}$$

where $\tilde{\theta}_{ij} = \theta_j - \hat{\theta}_{ij}$ and Γ_{ij} is a positive definite matrix. From (3.19) and (3.24), we obtain that

$$\dot{V}_2 = -\frac{k_1}{2}\delta^T Q\delta + \sum_{i=1}^{N}\Bigg[-k_{i,2}e_{i,2}^2 + e_{i,2}e_{i,3} + \tilde{\theta}_i^T\Gamma_i^{-1}\left(\Gamma_i\tau_{i,2} - \dot{\hat{\theta}}_i\right)$$

$$+e_{i,2}\frac{\partial\alpha_{i,1}}{\partial\hat{\theta}_i}\left(\Gamma_i\tau_{i,2} - \dot{\hat{\theta}}_i\right) + \sum_{j=1}^{N} a_{ij}\tilde{\theta}_{ij}^T\Gamma_{ij}^{-1}\left(\Gamma_{ij}\bar{\tau}_{ij,1} - \dot{\hat{\theta}}_{ij}\right)\Bigg], \tag{3.26}$$

where $\bar{\tau}_{ij,1}$ is defined as

$$\bar{\tau}_{ij,1} = -\frac{\partial\alpha_{i,1}}{\partial x_{j,1}}\varphi_{j,1}e_{i,2}, \quad \text{if } a_{ij} = 1. \tag{3.27}$$

Step3. We now clarify the arguments of $\alpha_{i,2}$. By examining (3.22) along with (3.15), (3.23), it can be seen that $\alpha_{i,2}$ is a function of $x_{i,1}$, $x_{i,2}$, $\hat{\theta}_i$, f_r, $\dot{f}_r$, $\ddot{f}_r$, $x_{j,1}$, $x_{j,2}$, $\hat{\theta}_{ij}$ (if $a_{ij} = 1$) and $\hat{\tilde{w}}_{ri}$ (if $\mu_i = 0$). Introduce a new error variable as

$$e_{i,4} = x_{i,4} - \alpha_{i,3}. \tag{3.28}$$

From (3.1), (3.22), and (3.28), the derivative of $e_{i,3}$ is computed as

$$\dot{e}_{i,3} = e_{i,4} + \alpha_{i,3} + \varphi_{i,3}^T\theta_i - \Bigg[\frac{\partial\alpha_{i,2}}{\partial x_{i,1}}\left(x_{i,2} + \varphi_{i,1}^T\theta_i\right) + \frac{\partial\alpha_{i,2}}{\partial x_{i,2}}\left(x_{i,3} + \varphi_{i,2}^T\theta_i\right)$$

$$+\frac{\partial\alpha_{i,2}}{\partial\hat{\theta}_i}\dot{\hat{\theta}}_i + \frac{\partial\alpha_{i,2}}{\partial f_r}\dot{f}_r + \frac{\partial\alpha_{i,2}}{\partial\dot{f}_r}\ddot{f}_r + \frac{\partial\alpha_{i,2}}{\partial\ddot{f}_r}f_r^{(3)} + (1-\mu_i)\frac{\partial\alpha_{i,2}}{\partial\hat{\tilde{w}}_{ri}}\dot{\hat{\tilde{w}}}_{ri}$$

$$+\sum_{j=1}^{N} a_{ij}\left(\frac{\partial\alpha_{i,2}}{\partial x_{j,1}}\left(x_{j,2} + \varphi_{j,1}^T\theta_j\right) + \frac{\partial\alpha_{i,2}}{\partial x_{j,2}}\left(x_{j,3} + \varphi_{j,2}^T\theta_j\right) + \frac{\partial\alpha_{i,2}}{\partial\hat{\theta}_{ij}}\dot{\hat{\theta}}_{ij}\right)\Bigg], \tag{3.29}$$

where $\alpha_{i,3}$ is chosen as

$$\alpha_{i,3} = -e_{i,2} - k_{i,3}e_{i,3} - \zeta_{i,3}^T\hat{\theta}_i + \frac{\partial\alpha_{i,2}}{\partial\hat{\theta}_i}\Gamma_i\tau_{i,3} + \frac{\partial\alpha_{i,2}}{\partial x_{i,1}}x_{i,2} + \frac{\partial\alpha_{i,2}}{\partial x_{i,2}}x_{i,3}$$

$$+\frac{\partial\alpha_{i,1}}{\partial\hat{\theta}_i}e_{i,2}\Gamma_i\zeta_{i,3} + \sum_{j=1}^{N} a_{ij}\Bigg[\frac{\partial\alpha_{i,2}}{\partial x_{j,1}}x_{j,2} + \frac{\partial\alpha_{i,2}}{\partial x_{j,2}}x_{j,3} + \zeta_{ij,2}^T\hat{\theta}_{ij}$$

$$+\frac{\partial\alpha_{i,2}}{\partial\hat{\theta}_{ij}}\Gamma_{ij}\bar{\tau}_{ij,2}\Bigg] + \sum_{l=1}^{3}\frac{\partial\alpha_{i,2}}{\partial f_r^{(l-1)}}f_r^{(l)} + (1-\mu_i)\frac{\partial\alpha_{i,2}}{\partial\hat{\tilde{w}}_{ri}}\dot{\hat{\tilde{w}}}_{ri}, \tag{3.30}$$

with $k_{i,3}$ a positive constant. $\zeta_{i,3}, \bar{\zeta}_{ij,2}, \tau_{i,3}, \bar{\tau}_{ij,2}$ are tuning functions defined as follows.

$$\zeta_{i,3} = \varphi_{i,3} - \frac{\partial \alpha_{i,2}}{\partial x_{i,1}} \varphi_{i,1} - \frac{\partial \alpha_{i,2}}{\partial x_{i,2}} \varphi_{i,2}, \tag{3.31}$$

$$\bar{\zeta}_{ij,2} = \frac{\partial \alpha_{i,2}}{\partial x_{j,1}} \varphi_{j,1} + \frac{\partial \alpha_{i,2}}{\partial x_{j,2}} \varphi_{j,2}, \quad \text{if } a_{ij} = 1, \tag{3.32}$$

$$\tau_{i,3} = \tau_{i,2} + \zeta_{i,3} e_{i,3}, \tag{3.33}$$

$$\bar{\tau}_{ij,2} = \bar{\tau}_{ij,1} - \bar{\zeta}_{ij,2} e_{i,3}, \quad \text{if } a_{ij} = 1. \tag{3.34}$$

From (3.20), (3.28)–(3.34), the derivative of $e_{i,3}$ is computed as

$$\dot{e}_{i,3} = -e_{i,2} - k_{i,3} e_{i,3} + e_{i,4} + \zeta_{i,3}^T \tilde{\theta}_i + \frac{\partial \alpha_{i,2}}{\partial \hat{\theta}_i} \left(\Gamma_i \tau_{i,3} - \dot{\hat{\theta}}_i \right) - \sum_{j=1}^{N} a_{ij} \bar{\zeta}_{ij,2}^T \tilde{\theta}_{ij}$$

$$+ \sum_{j=1}^{N} a_{ij} \frac{\partial \alpha_{i,2}}{\partial \hat{\theta}_{ij}} \left(\Gamma_{ij} \bar{\tau}_{ij,2} - \dot{\hat{\theta}}_{ij} \right) + \frac{\partial \alpha_{i,1}}{\partial \hat{\theta}_i} e_{i,2} \Gamma_i \zeta_{i,3}. \tag{3.35}$$

Define a Lyapunov function candidate V_3 at this step as

$$V_3 = V_2 + \frac{1}{2} \sum_{i=1}^{N} e_{i,3}^2. \tag{3.36}$$

From (3.26) and (3.35), we obtain that

$$\dot{V}_3 = -\frac{k_1}{2} \delta^T Q \delta + \sum_{i=1}^{N} \left[-k_{i,2} e_{i,2}^2 - k_{i,3} e_{i,3}^2 + e_{i,3} e_{i,4} + \tilde{\theta}_i^T \Gamma_i^{-1} \left(\Gamma_i \tau_{i,3} - \dot{\hat{\theta}}_i \right) \right.$$

$$+ \left(e_{i,3} \frac{\partial \alpha_{i,2}}{\partial \hat{\theta}_i} + e_{i,2} \frac{\partial \alpha_{i,1}}{\partial \hat{\theta}_i} \right) \left(\Gamma_i \tau_{i,3} - \dot{\hat{\theta}}_i \right)$$

$$+ \left. \sum_{j=1}^{N} a_{ij} \left(\tilde{\theta}_{ij}^T \Gamma_{ij}^{-1} \left(\Gamma_{ij} \bar{\tau}_{ij,2} - \dot{\hat{\theta}}_{ij} \right) + e_{i,3} \frac{\partial \alpha_{i,2}}{\partial \hat{\theta}_{ij}} \left(\Gamma_{ij} \bar{\tau}_{ij,2} - \dot{\hat{\theta}}_{ij} \right) \right) \right]. \tag{3.37}$$

Step 4. By examining (3.30), along with (3.15), (3.20), and (3.31)–(3.34), it can be seen that $\alpha_{i,3}$ is a function of $x_{i,1}, x_{i,2}, x_{i,3}, \hat{\theta}_i, f_r, \dot{f}_r, \ddot{f}_r, f_r^{(3)}, x_{j,1}, x_{j,2}, x_{j,3}, \hat{\theta}_{ij}$ (if $a_{ij} = 1$) and $\tilde{\bar{w}}_{ri}$ (if $\mu_i = 0$). Introduce a new error variable as

$$e_{i,5} = x_{i,5} - \alpha_{i,4}. \tag{3.38}$$

From (3.1), (3.30), and (3.38), the derivative of $e_{i,4}$ is computed as

$$\dot{e}_{i,4} = e_{i,5} + \alpha_{i,4} + \varphi_{i,4}^T \theta_i - \left[\sum_{l=1}^{3} \frac{\partial \alpha_{i,3}}{\partial x_{i,l}} \left(x_{i,l+1} + \varphi_{i,l}^T \theta_i \right) \right.$$

$$+\frac{\partial \alpha_{i,3}}{\partial \hat{\theta}_i}\dot{\hat{\theta}}_i + \sum_{l=1}^{4}\frac{\partial \alpha_{i,3}}{\partial f_r^{(l-1)}}f_r^{(l)} + (1-\mu_i)\frac{\partial \alpha_{i,3}}{\partial \hat{\tilde{w}}_{ri}}\dot{\hat{\tilde{w}}}_{ri}$$

$$+\sum_{j=1}^{N}a_{ij}\left(\sum_{l=1}^{3}\frac{\partial \alpha_{i,3}}{\partial x_{j,l}}(x_{j,l+1}+\varphi_{j,l}^T\theta_j)+\frac{\partial \alpha_{i,3}}{\partial \hat{\theta}_{ij}}\dot{\hat{\theta}}_{ij}\right)\Bigg],\qquad (3.39)$$

where $\alpha_{i,4}$ is chosen as

$$\alpha_{i,4}=-e_{i,3}-k_{i,4}e_{i,4}-\zeta_{i,4}^T\hat{\theta}_i+\frac{\partial \alpha_{i,3}}{\partial \hat{\theta}_i}\Gamma_i\tau_{i,4}+\sum_{l=1}^{3}\frac{\partial \alpha_{i,3}}{\partial x_{i,l}}x_{i,l+1}$$

$$+\left(\sum_{l=2}^{3}\frac{\partial \alpha_{i,l-1}}{\partial \hat{\theta}_i}e_{i,l}\right)\Gamma_i\zeta_{i,q}+\sum_{j=1}^{N}a_{ij}\left[\sum_{l=1}^{3}\frac{\partial \alpha_{i,3}}{\partial x_{j,l}}x_{j,l+1}+\bar{\zeta}_{ij,3}^T\hat{\theta}_{ij}\right.$$

$$+\frac{\partial \alpha_{i,3}}{\partial \hat{\theta}_{ij}}\Gamma_{ij}\bar{\tau}_{ij,3}-\frac{\partial \alpha_{i,2}}{\partial \hat{\theta}_{ij}}\Gamma_{ij}\bar{\zeta}_{ij,3}e_{i,3}\Bigg]+\sum_{l=1}^{4}\frac{\partial \alpha_{i,3}}{\partial f_r^{(l-1)}}f_r^{(l)}$$

$$+(1-\mu_i)\frac{\partial \alpha_{i,3}}{\partial \hat{\tilde{w}}_{ri}}\dot{\hat{\tilde{w}}}_{ri},\qquad (3.40)$$

with $k_{i,4}$ a positive constant. $\zeta_{i,4},\bar{\zeta}_{ij,3},\tau_{i,4},\bar{\tau}_{ij,3}$ are defined as follows.

$$\zeta_{i,4}=\varphi_{i,4}-\sum_{l=1}^{3}\frac{\partial \alpha_{i,3}}{\partial x_{i,l}}\varphi_{i,l},\qquad (3.41)$$

$$\bar{\zeta}_{ij,3}=\sum_{l=1}^{3}\frac{\partial \alpha_{i,3}}{\partial x_{j,l}}\varphi_{j,l},\qquad (3.42)$$

$$\tau_{i,4}=\tau_{i,3}+\zeta_{i,4}e_{i,4},\qquad (3.43)$$

$$\bar{\tau}_{ij,3}=\bar{\tau}_{ij,2}-\bar{\zeta}_{ij,3}e_{i,4},\quad \text{if}\quad a_{ij}=1.\qquad (3.44)$$

From (3.28), (3.38)–(3.44), the derivative of $e_{i,4}$ is computed as

$$\dot{e}_{i,4}=-e_{i,3}-k_{i,4}e_{i,4}+e_{i,5}+\zeta_{i,4}^T\tilde{\theta}_i+\frac{\partial \alpha_{i,3}}{\partial \hat{\theta}_i}\left(\Gamma_i\tau_{i,4}-\dot{\hat{\theta}}_i\right)$$

$$+\sum_{l=2}^{3}\frac{\partial \alpha_{i,l-1}}{\partial \hat{\theta}_i}e_{i,l}\Gamma_i\zeta_{i,3}+\sum_{j=1}^{N}a_{ij}\left[-\bar{\zeta}_{ij,3}^T\tilde{\theta}_{ij}+\frac{\partial \alpha_{i,3}}{\partial \hat{\theta}_{ij}}\left(\Gamma_{ij}\bar{\tau}_{ij,3}-\dot{\hat{\theta}}_{ij}\right)\right.$$

$$-\frac{\partial \alpha_{i,2}}{\partial \hat{\theta}_{ij}}\Gamma_{ij}\bar{\zeta}_{ij,3}e_{i,3}\Bigg].\qquad (3.45)$$

Define a Lyapunov function candidate V_4 at this step as

$$V_4=V_3+\frac{1}{2}\sum_{i=1}^{N}e_{i,4}^2.\qquad (3.46)$$

From (3.37) and (3.45), we obtain that

$$
\dot{V}_4 = - \frac{k_1}{2} \delta^T Q \delta + \sum_{i=1}^{N} \left[- \sum_{l=2}^{4} k_{i,l} e_{i,l}^2 + e_{i,4} e_{i,5} + \tilde{\theta}_i^T \Gamma_i^{-1} \left(\Gamma_i \tau_{i,4} - \dot{\hat{\theta}}_i \right) \right.
$$
$$
+ \left(\sum_{l=2}^{4} e_{i,l} \frac{\partial \alpha_{i,l-1}}{\partial \hat{\theta}_i} \right) \left(\Gamma_i \tau_{i,4} - \dot{\hat{\theta}}_i \right) + \sum_{j=1}^{N} a_{ij} \tilde{\theta}_{ij}^T \Gamma_{ij}^{-1} \left(\Gamma_{ij} \bar{\tau}_{ij,3} - \dot{\hat{\theta}}_{ij} \right)
$$
$$
\left. + \sum_{j=1}^{N} a_{ij} \left(\sum_{l=3}^{4} e_{i,l} \frac{\partial \alpha_{i,l-1}}{\partial \hat{\theta}_{ij}} \right) \left(\Gamma_{ij} \bar{\tau}_{ij,3} - \dot{\hat{\theta}}_{ij} \right) \right], \tag{3.47}
$$

Step q ($q = 4, \dots, n-1$) It can be seen that $\alpha_{i,q-1}$ is a function of $x_{i,1}, \cdots, x_{i,q-1}$, $\hat{\theta}_i, f_r, \cdots, f_r^{q-1}, x_{j,1}, \cdots, x_{j,q-1}, \hat{\theta}_{ij}$ (if $a_{ij} = 1$) and $\hat{\bar{w}}_{ri}$ (if $\mu_i = 0$) from Step 1 to Step $q-1$. Introduce a new error variable as

$$
e_{i,q+1} = x_{i,q+1} - \alpha_{i,q}. \tag{3.48}
$$

From (3.1) and (3.48), the derivative of $e_{i,q}$ is computed as

$$
\dot{e}_{i,q} = e_{i,q+1} + \alpha_{i,q} + \varphi_{i,q}^T \theta_i - \left[\sum_{l=1}^{q-1} \frac{\partial \alpha_{i,q-1}}{\partial x_{i,l}} \left(x_{i,l+1} + \varphi_{i,l}^T \theta_i \right) \right.
$$
$$
+ \frac{\partial \alpha_{i,q-1}}{\partial \hat{\theta}_i} \dot{\hat{\theta}}_i + \sum_{l=1}^{q} \frac{\partial \alpha_{i,q-1}}{\partial f_r^{(l-1)}} f_r^{(l)} + (1 - \mu_i) \frac{\partial \alpha_{i,q-1}}{\partial \hat{\bar{w}}_{ri}} \dot{\hat{\bar{w}}}_{ri}
$$
$$
\left. + \sum_{j=1}^{N} a_{ij} \left(\sum_{l=1}^{q-1} \frac{\partial \alpha_{i,q-1}}{\partial x_{j,l}} \left(x_{j,l+1} + \varphi_{j,l}^T \theta_j \right) + \frac{\partial \alpha_{i,q-1}}{\partial \hat{\theta}_{ij}} \dot{\hat{\theta}}_{ij} \right) \right] \tag{3.49}
$$

where $\alpha_{i,q}$ is chosen as

$$
\alpha_{i,q} = - e_{i,q-1} - k_{i,q} e_{i,q} - \zeta_{i,q}^T \hat{\theta}_i + \frac{\partial \alpha_{i,q-1}}{\partial \hat{\theta}_i} \Gamma_i \tau_{i,q} + \sum_{l=1}^{q-1} \frac{\partial \alpha_{i,q-1}}{\partial x_{i,l}} x_{i,l+1}
$$
$$
+ \left(\sum_{l=2}^{q-1} \frac{\partial \alpha_{i,l-1}}{\partial \hat{\theta}_i} e_{i,l} \right) \Gamma_i \zeta_{i,q} + \sum_{j=1}^{N} a_{ij} \left[\sum_{l=1}^{q-1} \frac{\partial \alpha_{i,q-1}}{\partial x_{j,l}} x_{j,l+1} + \bar{\zeta}_{ij,q-1}^T \hat{\theta}_{ij} \right.
$$
$$
\left. + \frac{\partial \alpha_{i,q-1}}{\partial \hat{\theta}_{ij}} \Gamma_{ij} \bar{\tau}_{ij,q-1} - \sum_{l=3}^{q-1} \frac{\partial \alpha_{i,l-1}}{\partial \hat{\theta}_{ij}} \Gamma_{ij} \bar{\zeta}_{ij,q-1} e_{i,l} \right]
$$
$$
+ \sum_{l=1}^{q} \frac{\partial \alpha_{i,q-1}}{\partial f_r^{(l-1)}} f_r^{(l)} + (1 - \mu_i) \frac{\partial \alpha_{i,q-1}}{\partial \hat{\bar{w}}_{ri}} \dot{\hat{\bar{w}}}_{ri}, \tag{3.50}
$$

with $k_{i,q}$ a positive constant. $\zeta_{i,q}, \bar{\zeta}_{ij,q-1}, \tau_{i,q}, \bar{\tau}_{ij,q-1}$ are defined as follows.

$$
\zeta_{i,q} = \varphi_{i,q} - \sum_{l=1}^{q-1} \frac{\partial \alpha_{i,q-1}}{\partial x_{i,l}} \varphi_{i,l}, \tag{3.51}
$$

$$\bar{\zeta}_{ij,q-1} = \sum_{l=1}^{q-1} \frac{\partial \alpha_{i,q-1}}{\partial x_{j,l}} \varphi_{j,l}, \quad \text{if} \quad a_{ij} = 1, \tag{3.52}$$

$$\tau_{i,q} = \tau_{i,q-1} + \zeta_{i,q} e_{i,q}, \tag{3.53}$$

$$\bar{\tau}_{ij,q-1} = \bar{\tau}_{ij,q-2} - \bar{\zeta}_{ij,q-1} e_{i,q}, \quad \text{if} \quad a_{ij} = 1. \tag{3.54}$$

From (3.48)-(3.54), the derivative of $e_{i,q}$ is computed as

$$
\begin{aligned}
\dot{e}_{i,q} = {}& - e_{i,q-1} - k_{i,q} e_{i,q} + e_{i,q+1} + \zeta_{i,q}^T \tilde{\theta}_i + \frac{\partial \alpha_{i,q-1}}{\partial \hat{\theta}_i} \left(\Gamma_i \tau_{i,q} - \dot{\hat{\theta}}_i \right) \\
& + \sum_{l=2}^{q-1} \frac{\partial \alpha_{i,l-1}}{\partial \hat{\theta}_i} e_{i,l} \Gamma_i \zeta_{i,q-1} + \sum_{j=1}^{N} a_{ij} \Bigg[- \bar{\zeta}_{ij,q-1}^T \tilde{\theta}_{ij} \\
& + \frac{\partial \alpha_{i,q-1}}{\partial \hat{\theta}_{ij}} \left(\Gamma_{ij} \bar{\tau}_{ij,q-1} - \dot{\hat{\theta}}_{ij} \right) - \sum_{l=3}^{q-1} \left(\frac{\partial \alpha_{i,l-1}}{\partial \hat{\theta}_{ij}} e_{i,l} \right) \Gamma_{ij} \bar{\zeta}_{ij,q-1} \Bigg] \tag{3.55}
\end{aligned}
$$

Define a Lyapunov function candidate V_q at this step as

$$V_q = V_{q-1} + \frac{1}{2} \sum_{i=1}^{N} e_{i,q}^2. \tag{3.56}$$

From (3.37) and (3.55), we obtain that

$$
\begin{aligned}
\dot{V}_q = {}& - \frac{k_1}{2} \delta^T Q \delta + \sum_{i=1}^{N} \Bigg[- \sum_{l=2}^{q} k_{i,l} e_{i,l}^2 + e_{i,q} e_{i,q+1} + \tilde{\theta}_i^T \Gamma_i^{-1} \left(\Gamma_i \tau_{i,q} - \dot{\hat{\theta}}_i \right) \\
& + \left(\sum_{l=2}^{q} e_{i,l} \frac{\partial \alpha_{i,l-1}}{\partial \hat{\theta}_i} \right) \left(\Gamma_i \tau_{i,q} - \dot{\hat{\theta}}_i \right) + \sum_{j=1}^{N} a_{ij} \tilde{\theta}_{ij}^T \Gamma_{ij}^{-1} \left(\Gamma_{ij} \bar{\tau}_{ij,q-1} - \dot{\hat{\theta}}_{ij} \right) \\
& + \sum_{j=1}^{N} a_{ij} \left(\sum_{l=3}^{q} e_{i,l} \frac{\partial \alpha_{i,l-1}}{\partial \hat{\theta}_{ij}} \right) \left(\Gamma_{ij} \bar{\tau}_{ij,q-1} - \dot{\hat{\theta}}_{ij} \right) \Bigg]. \tag{3.57}
\end{aligned}
$$

Step n. It can be seen that $\alpha_{i,n-1}$ is a function of $x_{i,1}, \cdots, x_{i,n-1}$, $\hat{\theta}_i$, $f_r, \cdots, f_r^{(n-1)}$, $x_{j,1}, \cdots, x_{j,n-1}, \hat{\theta}_{ij}$ (if $a_{ij} = 1$) and $\hat{\bar{w}}_{ri}$ (if $\mu_i = 0$). From (3.1), (3.50), and (3.48), the derivative of $e_{i,n}$ is computed as

$$
\begin{aligned}
\dot{e}_{i,n} = {}& b_i \beta_i(x_i) u_i + \varphi_{i,n}^T \theta_i - \Bigg[\sum_{l=1}^{n-1} \frac{\partial \alpha_{i,n-1}}{\partial x_{i,l}} \left(x_{i,l+1} + \varphi_{i,l}^T \theta_i \right) \\
& + \frac{\partial \alpha_{i,n-1}}{\partial \hat{\theta}_i} \dot{\hat{\theta}}_i + \sum_{l=1}^{n} \frac{\partial \alpha_{i,n-1}}{\partial f_r^{(l-1)}} f_r^{(l)} + (1 - \mu_i) \frac{\partial \alpha_{i,n-1}}{\partial \hat{\bar{w}}_{ri}} \dot{\hat{\bar{w}}}_{ri} \\
& + \sum_{j=1}^{N} a_{ij} \left(\sum_{l=1}^{n-1} \frac{\partial \alpha_{i,n-1}}{\partial x_{j,l}} \left(x_{j,l+1} + \varphi_{j,l}^T \theta_j \right) + \frac{\partial \alpha_{i,n-1}}{\partial \hat{\theta}_{ij}} \dot{\hat{\theta}}_{ij} \right) \Bigg], \tag{3.58}
\end{aligned}
$$

where $\alpha_{i,n}$ is chosen as

$$
\alpha_{i,n} = - e_{i,n-1} - k_{i,n}e_{i,n} - \zeta_{i,n}^T \hat{\theta}_i + \frac{\partial \alpha_{i,n-1}}{\partial \hat{\theta}_i}\Gamma_i\tau_{i,n} + \sum_{l=1}^{n-1} \frac{\partial \alpha_{i,n-1}}{\partial x_{i,l}}x_{i,l+1}
$$

$$
+ \left(\sum_{l=2}^{n-1} \frac{\partial \alpha_{i,l-1}}{\partial \hat{\theta}_i} e_{i,l} \right)\Gamma_i\zeta_{i,n} + \sum_{j=1}^{N} a_{ij}\left[\sum_{l=1}^{n-1} \frac{\partial \alpha_{i,n-1}}{\partial x_{j,l}}x_{j,l+1} \right.
$$

$$
\left. + \bar{\zeta}_{ij,n-1}^T \hat{\theta}_{ij} + \frac{\partial \alpha_{i,n-1}}{\partial \hat{\theta}_{ij}}\Gamma_{ij}\bar{\tau}_{ij,n-1} - \sum_{l=3}^{n-1} \frac{\partial \alpha_{i,l-1}}{\partial \hat{\theta}_{ij}}\Gamma_{ij}\bar{\zeta}_{ij,n-1}e_{i,l} \right]
$$

$$
+ \sum_{l=1}^{n} \frac{\partial \alpha_{i,n-1}}{\partial f_r^{(l-1)}}f_r^{(l)} + (1 - \mu_i)\frac{\partial \alpha_{i,n-1}}{\partial \hat{w}_{ri}}\dot{\hat{w}}_{ri}, \tag{3.59}
$$

with $k_{i,n}$ a positive constant. $\zeta_{i,n}, \bar{\zeta}_{ij,n-1}, \tau_{i,n}, \bar{\tau}_{ij,n-1}$ are defined as follows.

$$
\zeta_{i,n} = \varphi_{i,n} - \sum_{l=1}^{n-1} \frac{\partial \alpha_{i,n-1}}{\partial x_{i,l}}\varphi_{i,l}, \tag{3.60}
$$

$$
\bar{\zeta}_{ij,n-1} = \sum_{l=1}^{n-1} \frac{\partial \alpha_{i,n-1}}{\partial x_{j,l}}\varphi_{j,l}, \quad \text{if} \quad a_{ij} = 1, \tag{3.61}
$$

$$
\tau_{i,n} = \tau_{i,n-1} + \zeta_{i,n}e_{i,n}, \tag{3.62}
$$

$$
\bar{\tau}_{ij,n-1} = \bar{\tau}_{ij,n-2} - \bar{\zeta}_{ij,n-1}e_{i,n}, \quad \text{if} \quad a_{ij} = 1. \tag{3.63}
$$

The control law and the parameter estimators are designed as

$$
u_i = \frac{\hat{\varrho}_i}{\beta_i(x_i)}\alpha_{i,n}, \tag{3.64}
$$

$$
\dot{\hat{\varrho}}_i = -\gamma_i\text{sgn}(b_i)\alpha_{i,n}e_{i,n}, \tag{3.65}
$$

$$
\dot{\hat{\theta}}_i = \Gamma_i\tau_{i,n}, \tag{3.66}
$$

$$
\dot{\hat{\theta}}_{ij} = \Gamma_{ij}\bar{\tau}_{i,n-1}, \tag{3.67}
$$

where $\hat{\varrho}_i$ is the estimate of $\varrho = 1/b_i$.

For easier reading, the design details of the above steps are summarized in Table 3.1.

3.4 STABILITY AND CONSENSUS ANALYSIS

The main results of our distributed adaptive control design scheme can be formally stated in the following theorem.

Theorem 3.1 *Consider the closed-loop adaptive system consisting of N uncertain nonlinear agents (3.1) satisfying Assumptions 3.1.1–3.1.3, the local controllers*

TABLE 3.1

The Design of Distributed Adaptive Controllers for Step q ($q = 4, \ldots, n$).

Introduce Error Variables:

$$e_{i,q+1} = x_{i,q+1} - \alpha_{i,q}. \tag{3.68}$$

Control Laws:

$$u_i = \frac{\hat{\varrho}_i}{\beta_i(x_i)} \alpha_{i,n}, \tag{3.69}$$

with

$$
\begin{aligned}
\alpha_{i,q} &= -e_{i,q-1} - k_{i,q}e_{i,q} - \zeta_{i,q}^T \hat{\theta}_i + \frac{\partial \alpha_{i,q-1}}{\partial \hat{\theta}_i} \Gamma_i \tau_{i,q} \\
&\quad + \sum_{l=1}^{q-1} \frac{\partial \alpha_{i,q-1}}{\partial x_{i,l}} x_{i,l+1} + \left(\sum_{l=2}^{q-1} \frac{\partial \alpha_{i,l-1}}{\partial \hat{\theta}_i} e_{i,l} \right) \Gamma_i \zeta_{i,q} \\
&\quad + \sum_{j=1}^{N} a_{ij} \left[\sum_{l=1}^{q-1} \frac{\partial \alpha_{i,q-1}}{\partial x_{j,l}} x_{j,l+1} + \bar{\zeta}_{ij,q-1}^T \hat{\theta}_{ij} \right. \\
&\quad + \frac{\partial \alpha_{i,q-1}}{\partial \hat{\theta}_{ij}} \Gamma_{ij} \bar{\tau}_{ij,q-1} - \left. \sum_{l=3}^{q-1} \frac{\partial \alpha_{i,l-1}}{\partial \hat{\theta}_{ij}} \Gamma_{ij} \bar{\zeta}_{ij,q-1} e_{i,l} \right] \\
&\quad + \sum_{l=1}^{q} \frac{\partial \alpha_{i,q-1}}{\partial f_r^{(l-1)}} f_r^{(l)} + (1 - \mu_i) \frac{\partial \alpha_{i,q-1}}{\partial \hat{w}_{ri}} \dot{\hat{w}}_{ri}, \tag{3.70}
\end{aligned}
$$

$$\zeta_{i,q} = \varphi_{i,q} - \sum_{l=1}^{q-1} \frac{\partial \alpha_{i,q-1}}{\partial x_{i,l}} \varphi_{i,l}, \tag{3.71}$$

$$\bar{\zeta}_{ij,q-1} = \sum_{l=1}^{q-1} \frac{\partial \alpha_{i,q-1}}{\partial x_{j,l}} \varphi_{j,l}, \tag{3.72}$$

$$\tau_{i,q} = \tau_{i,q-1} + \zeta_{i,q} e_{i,q}, \tag{3.73}$$

$$\bar{\tau}_{ij,q-1} = \bar{\tau}_{ij,q-2} - \bar{\zeta}_{ij,q-1} e_{i,q}. \tag{3.74}$$

Parameter Estimators:

$$\dot{\hat{\varrho}}_i = -\gamma_i \mathrm{sgn}(b_i) \alpha_{i,n} e_{i,n}, \tag{3.75}$$

$$\dot{\hat{\theta}}_i = \Gamma_i \tau_{i,n}, \tag{3.76}$$

$$\dot{\hat{\theta}}_{ij} = \Gamma_{ij} \bar{\tau}_{i,n-1}. \tag{3.77}$$

(3.69) and the parameter estimators (3.17), (3.75)–(3.77). All the signals in the closed-loop system are globally uniformly bounded and asymptotic consensus tracking of all the agents' outputs to $y_r(t)$ is achieved, i.e., $\lim_{t\to\infty}[y_i(t)-y_r(t)] = 0$ for $i = 1,\ldots,N$.

Proof. We define the Lyapunov function for the overall system as

$$V_n = V_3 + \frac{1}{2}\sum_{i=1}^{N}\left(\sum_{q=4}^{n}e_{i,q}^2 + \frac{|b_i|}{\gamma_i}\tilde{\varrho}_i^2\right), \tag{3.78}$$

where $\tilde{\varrho}_i = \varrho_i - \hat{\varrho}_i$. From (3.37) to (3.74), the derivative of V_n is computed as

$$
\begin{aligned}
\dot{V}_n &= -\frac{k_1}{2}\delta^T Q\delta + \sum_{i=1}^{N}\left[-\sum_{l=2}^{n}k_{i,l}e_{i,l}^2 + \tilde{\theta}_i^T\Gamma_i^{-1}\left(\Gamma_i\tau_{i,n} - \dot{\hat{\theta}}_i\right)\right.\\
&\quad + \left(\sum_{l=2}^{n}e_{i,l}\frac{\partial\alpha_{i,l-1}}{\partial\hat{\theta}_i}\right)\left(\Gamma_i\tau_{i,n} - \dot{\hat{\theta}}_i\right) + \sum_{j=1}^{N}a_{ij}\tilde{\theta}_{ij}^T\Gamma_{ij}^{-1}\left(\Gamma_{ij}\bar{\tau}_{ij,n-1} - \dot{\hat{\theta}}_{ij}\right)\\
&\quad + \sum_{j=1}^{N}a_{ij}\left(\sum_{l=3}^{n}e_{i,l}\frac{\partial\alpha_{i,l-1}}{\partial\hat{\theta}_{ij}}\right)\left(\Gamma_{ij}\bar{\tau}_{ij,n-1} - \dot{\hat{\theta}}_{ij}\right)\\
&\quad \left. + \frac{|b_i|}{\gamma_i}\tilde{\varrho}_i\left(-\dot{\hat{\varrho}}_i - \gamma_i\mathrm{sgn}(b_i)\alpha_{i,n}\right)\right].
\end{aligned}
\tag{3.79}
$$

Based on Assumption 3.1.3 and Lemma 2.4, Q is positive definite. Thus by choosing the parameter update laws as (3.75)–(3.77), $\dot{V}_n$ can be rendered negative semi-definite such that

$$\dot{V}_n = -\frac{k}{2}\delta^T Q\delta - \sum_{i=1}^{N}\sum_{k=2}^{n}k_{i,l}e_{i,l}^2. \tag{3.80}$$

From the definition of V_n in (3.78) along with (3.18), (3.25), and (3.36), we establish that $e_{i,l}$ for $l = 1,\ldots,n$, $\hat{\theta}_i$, $\hat{\theta}_{ij}$, $\hat{\varrho}_i$, and $\hat{\tilde{w}}_{ri}$ are bounded for all agent i. From (3.4) and the boundedness of f_r given in Assumption 3.1.1, we obtain that y_i, i.e., $x_{i,1}$ for $i = 1,\ldots,N$ are bounded. From (3.6) and $\delta = y-\underline{y}_r$, z is also bounded. From (3.16) and the smoothness of φ_i, $\alpha_{i,1}$ for $i = 1,\ldots,N$ are bounded. From the definition of $e_{i,2}$ in (3.15), it follows that $x_{i,2}$ is bounded. By following similar procedure, the boundedness of $\alpha_{i,q}$ and $x_{i,q}$ for $q = 3,\ldots,n$ is ensured. From (3.69), we can conclude that the control signal u_i is bounded. Thus the boundedness of all the signals in the closed-loop adaptive systems is guaranteed. By applying the LaSalle-Yoshizawa theorem, it further follows that $\lim_{t\to\infty}\delta_i(t) = 0$ for $i = 1,\ldots,N$ and $q = 1,\ldots,n$. This implies that asymptotic consensus tracking of all the N agents' outputs to a desired trajectory $y_r(t)$ is also achieved, i.e., $\lim_{t\to\infty}[y_i(t) - y_r(t)] = 0$ for $i = 1,\ldots,N$. $\qquad\square$

Remark 3.5 *The distributed adaptive control scheme presented in this chapter is also applicable to the following two cases if there is at least one agent in $\mathcal{G}$ has direct* *access to y_r. i) The graph $\mathcal{G}$ is undirected and connected; ii) The graph $\mathcal{G}$ is directed, balanced, and strongly connected. This is because under these two cases, matrices $(\mathcal{L} + \mathcal{B})$ and $(\mathcal{L} + \mathcal{L}^T + 2\mathcal{B})$ are symmetric positive definite, respectively [112, 161]. Thus by modifying $\alpha_{i,1}$ in (3.16) with P_i chosen as $P_i = 1$, it can be shown that the control objective is achieved by following similar analysis in the proof of Theorem 3.1.*

Remark 3.6 *Similar to [139], in constructing the local controller u_i with our proposed method as presented in Steps 1 and 2 and Table 1, $\varphi_{j,l}$ involving the structural knowledge of its neighbors' intrinsic dynamics needs to be collected if $a_{ij} = 1$. Moreover, we assume that the states x_j for agents $j \in \mathcal{N}_i$ is available for agent i at each time instant which is also required in [139]. However, in contrast to [139], by introducing $\hat{\theta}_{ij}$ in agent i to estimate the uncertain parameters (i.e., θ_j) contained in its neighbors' dynamics, information exchange of local parameter estimates among linked agents is avoided and the communication burden can thus be reduced. Moreover, no other conditions such as LMI are required to achieve the main results than Assumptions 3.1.1–3.1.3 which are checkable and mild. Besides, the parameter update laws are designed in a totally distributed manner without further information exchange of local synchronization errors among agents as required in [155].*

Remark 3.7 *The idea of introducing local estimators for unknown trajectory parameters is analogous to [6], in which the adaptive controllers are designed based on passivity framework [2]. Similar to [6], the consensus tracking is achieved directly no matter whether $\hat{\bar{w}}_{ri}$ will converge to the true values. Nevertheless, the coordination method presented in [6] is not applicable to directed graph as in our chapter since the symmetry property of undirected graphs cannot hold for directed graphs even when the graphs are balanced and strongly connected.*

3.5 SIMULATION RESULTS

We now consider an example to illustrate the proposed design schemes and verify the established theoretical results. Suppose that there is a group of four nonlinear agents with the following dynamics

$$\dot{x}_{i,1} = x_{i,2} + \varphi_{i,1}(x_{i,1})\theta_i,$$
$$\dot{x}_{i,2} = b_i u_i + \varphi_{i,2}(x_{i,1}, x_{i,2})\theta_i, \qquad i = 1, \ldots, 4, \qquad (3.81)$$

where $\varphi_{i,1} = \sin(x_{i,1})$, $\varphi_{1,2} = x_{1,2}^3$, $\varphi_{2,2} = x_{2,2}^2$, $\varphi_{3,2} = x_{3,2}$, $\varphi_{4,2} = x_{4,1}x_{4,2}$; $\theta_1 = 1$, $\theta_2 = 0.5$, $\theta_3 = -2$, $\theta_4 = -3$; $b_1 = 1$, $b_2 = -2$, $b_3 = 0.5$, $b_4 = 3$. The graph for these four agents is given in Figure 3.1. The desired trajectory is given by $y_r(t) = \sin(t)$. In simulation, all the state initials are set as zero except that $x_{1,1}(0) = 1$, $x_{2,1}(0) = 0.5$, $x_{4,1}(0) = -0.9$, and $x_{5,1}(0) = -1.2$. Besides, the design parameters are chosen as $k_1 = 2$, $k_{i,2} = 1$, $\gamma_i = \Gamma_i = \Gamma_{ri} = 1$ for $1 \leq i \leq 4$. The adaptive gains $\Gamma_{21} = \Gamma_{23} = \Gamma_{32} = \Gamma_{41} = 1$. The tracking errors ($\delta_i$) for all the

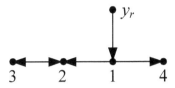

FIGURE 3.1 Graph for a group of four nonlinear agents.

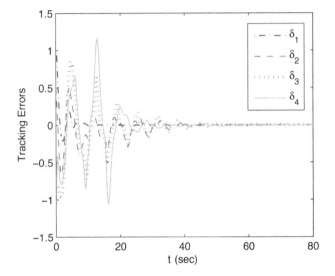

FIGURE 3.2 Tracking errors $\delta_i = y_i - y_r$ for $1 \leq i \leq 4$.

agents are shown in Figure 3.2. It can be seen that asymptotically consensus tracking is achieved with the proposed distributed adaptive control scheme.

3.6 NOTES

In this chapter, we have investigated the output consensus tracking control problem for a collection of nonlinear subsystems with intrinsic mismatched unknown parameters. Only part of the subsystems can obtain the exact information of the desired trajectory and the graph is represented by a directed graph. By adopting backstepping technique, distributed adaptive control laws are designed based on the information collected within neighboring areas. It is shown that all signals in the closed-loop system are bounded and the asymptotically consensus tracking for all the subsystems' outputs can be ensured. The simulation results show the effectiveness of the proposed control approach. Moreover, in contrast to currently available schemes in which information exchanges of online parameter estimates or local consensus errors among linked subsystems are required, the conditions can be relaxed by introducing additional local estimates to account for the uncertainties in reference signals and the neighbors' dynamics. Thus the desired results are achieved with less transmission burden.

4 Auxiliary Filters-Based Distributed Adaptive Consensus Control

In Chapter 3, the distributed adaptive consensus tracking control problem for high-order nonlinear multi-agent systems with mismatched unknown parameters and time-varying desired trajectory under directed graph has been studied. A novel distributed adaptive consensus tracking control scheme is proposed. However, we make the assumption that the nonlinear functions of the agents can be linearly parameterized with a set of basis functions that are known to all agents. While this is generally not a strong assumption, as elaborated in Remark 3.1, there is room for refinement.

In this chapter, we revisit the same problem and intend to remove the above assumption. Besides, undesired external disturbances are also considered. It is assumed that the desired trajectory $y_r(t)$ is known to only part of the agents in the group. By introducing local estimators for the bounds of reference trajectory and a filter for each agent, a new backstepping-based smooth distributed adaptive control protocol is proposed. Compared with the results in Chapter 3 and related results in the literature, the main advantages of the consensus control scheme proposed in this chapter can be summarized as follows. i) The non-smooth signum function based distributed control approaches in [26, 64, 75, 77] are undesired due to chattering phenomenon. To address this issue, new compensating terms are introduced including some smooth functions of consensus errors and a positive integrable time-varying signal, with which smooth consensus controllers are obtained. Furthermore, different from [30, 153, 161], global uniform boundedness of all the closed-loop signals and asymptotically consensus tracking for all agent outputs are achieved in this chapter. ii) In contrast to [6, 43, 130, 155], the assumptions of linearly parameterized reference signals and the corresponding basis function vectors being known to all agents are no longer needed. iii) The considered multi-agent system models are more general than those in most of the existing results on distributed consensus control including [2, 6, 16, 40, 105].

4.1 PROBLEM FORMULATION

We consider a group of N nonlinear agents which can be modeled as follows.

$$
\begin{aligned}
\dot{x}_{i,q} &= x_{i,q+1} + \varphi_{i,q}(x_{i,1}, \ldots, x_{i,q})^T \theta_i, \quad q = 1, \ldots, n-1, \\
\dot{x}_{i,n} &= u_i + \varphi_{i,n}(x_i)^T \theta_i + d_i(t), \\
y_i &= x_{i,1}, \quad \text{for } i = 1, 2, \ldots, N,
\end{aligned} \tag{4.1}
$$

DOI: 10.1201/9781003394372-4

where $x_i = [x_{i,1}, \ldots, x_{i,n}]^T \in \Re^n$, $u_i \in \Re$ and $y_i \in \Re$ are the states, control input and output of the ith agent, respectively. $\theta_i \in \Re^{p_i}$ is a vector of unknown constants. $\varphi_{i,j} : \Re^j \to \Re^{p_i}$ for $j = 1, \ldots, n$ are known smooth nonlinear functions. $d_i(t)$ represents the external disturbance. The desired trajectory is generated by a time-varying function $y_r(t)$.

We now use $\mu_i = 1$ to indicate the case that $y_r(t)$ is accessible directly to agent i; otherwise, μ_i is set as $\mu_i = 0$. The *control objective* is, while only part of the followers have access to the leader, to design distributed adaptive smooth controllers u_i for each agent by utilizing only locally available information obtained from the intrinsic agent and its neighbors such that:

- All the signals in the closed-loop system are globally uniformly bounded;
- The outputs of all the overall systems can track the desired trajectory $y_r(t)$ asymptotically, i.e., $\lim_{t \to \infty} [y_i(t) - y_r(t)] = 0$, $\forall i \in \mathcal{V}$.

To achieve the objective, the following assumptions are imposed.

Assumption 4.1.1 *The first nth-order derivatives of $y_r(t)$ are bounded, piecewise continuous. Let F_j, $j = 1, \ldots, n$ being the bound of jth-order derivative of $y_r(t)$, then F_j is also available to agent i if $\mu_i = 1$.*

Assumption 4.1.2 *The directed graph $\mathcal{G}$ contains a spanning tree with the root agent being the leader.*

Remark 4.1 *The multi-agent system model described in (4.1) is more general than those in most of the currently available results on distributed consensus control including [2, 6, 16, 40, 155] for the following terms. i) The agents are nonlinear and allowed to have arbitrary relative degree and non-identical dynamics; ii) intrinsic mismatched unknown parameters and uncertain disturbances are simultaneously involved.*

Remark 4.2 *Assumption 4.1.1 indicates that the bounds of up to nth-order derivatives of $y_r(t)$ are available to agent i if $\mu_i = 1$. This is a mild assumption because it is a common case that part of the agents may have full knowledge of the reference trajectory $y_r(t)$.*

4.2 DESIGN OF DISTRIBUTED FILTERS

Firstly, for each agent, a filter $(q_{i,1}, \ldots, q_{i,n})$ is designed. Design filter $q_{i,j}$ for agent i, $i = 1, \ldots, N$, $j = 1, \ldots, n$ as

$$\begin{cases} \dot{q}_{i,1} = q_{i,2}, \\ \vdots \\ \dot{q}_{i,n} = v_i, \end{cases} \tag{4.2}$$

and introduce the error variables

$$z_{i,j} = \sum_{k=1}^{N} a_{ik}(q_{i,j} - q_{k,j}) + \mu_i(q_{i,j} - y_r^{(j-1)}),$$ (4.3)

where $y_r^{(j-1)} = \frac{d^{(j-1)} y_r}{dt^j}$ and define $\hat{F}_{ij}$ for agent i to estimate the bound of $\frac{d^{(j)} y_r}{dt^j}$, i.e., F_j, as

$$\dot{\hat{F}}_{ij} = \sum_{k=1}^{N} a_{ik}(\hat{F}_{kj} - \hat{F}_{ij}) + \mu_i(F_j - \hat{F}_{ij}).$$ (4.4)

Let $z_j = [z_{1,j}, ..., z_{N,j}]^T$ and $q_j = [q_{1,j}, ..., q_{N,j}]^T$, then

$$z_j = (\mathcal{L} + \mathcal{B})(q_j - \underline{y}_r^{(j-1)}),$$ (4.5)

where $\underline{y}_r^{(j)} = [y_r^{(j)}, ..., y_r^{(j)}]^T$. Define

$$z = \left(\lambda + \frac{d}{dt}\right)^{n-1} z_1$$
$$= C_{n-1}^0 \lambda^{n-1} z_1 + C_{n-1}^1 \lambda^{n-2} z_2 + ... + C_{n-1}^{n-1} z_n,$$ (4.6)

and let $z = [\underline{z}_1, ..., \underline{z}_n]^T$, then

$$\dot{z} = (\mathcal{L} + \mathcal{B})(C_{n-1}^0 \lambda^{n-1} z_2 + C_{n-1}^1 \lambda^{n-2} z_3 + ...$$
$$+ C_{n-1}^{n-2} \lambda z_n + v - \sum_{j=1}^{n} \delta_j \underline{y}_r^{(j)}),$$ (4.7)

where $\delta_j = C_{n-1}^{j-1} \lambda^{n-j}$, $v = [v_1, ..., v_N]^T$. The distributed control for v_i as

$$v_i = - c\underline{z}_i - C_{n-1}^0 \lambda^{n-1} z_{i,2} - C_{n-1}^1 \lambda^{n-2} z_{i,3} - ...$$
$$- C_{n-1}^{n-2} \lambda z_{i,n} - \sum_{j=1}^{n} \frac{\delta_j |z_i| z_i}{z_i^2 + \varepsilon(t)^2} \hat{F}_{ij},$$ (4.8)

where $\varepsilon(t) = \beta e^{-\alpha t}$, α, β, and c are positive constants. Define $\hat{F}_j = [\hat{F}_{1j}, ..., \hat{F}_{Nj}]^T$ and $\underline{F}_j = [F_j, ..., F_j]^T \in \Re^N$, then

$$\dot{\tilde{F}}_j = -(\mathcal{L} + \mathcal{B})\tilde{F}_j,$$ (4.9)

where $\tilde{F}_j = [\tilde{F}_{1j}, ..., \tilde{F}_{Nj}]^T = \hat{F}_j - \underline{F}_j$ for $j = 1, ..., n$. Let $\epsilon(\underline{z}_i) = \text{sgn}(\underline{z}_i) - \frac{|z_i| z_i}{z_i^2 + \varepsilon(t)^2}$, then

$$\dot{z} = (\mathcal{L} + \mathcal{B})(-cz - \sum_{j=1}^{n} \delta_j \underline{\text{sgn}}(z).*\underline{F}_j - \sum_{j=1}^{n} \delta_j \underline{y}_r^{(j)})$$

$$+ \sum_{j=1}^{n} \delta_j \epsilon(z). * \underline{F}_j - \sum_{j=1}^{n} \delta_j Z. * \tilde{F}_j), \qquad (4.10)$$

where $Z = [\frac{|z_1| z_1}{z_1^2 + \varepsilon(t)^2}, ..., \frac{|z_N| z_N}{z_N^2 + \varepsilon(t)^2}]^T$, $\epsilon(z) = [\epsilon(\underline{z}_1), ..., \epsilon(\underline{z}_N)]^T$, and $\underline{\text{sgn}}(z) = [\text{sgn}(\underline{z}_1), ..., \text{sgn}(\underline{z}_N)]^T$.

Remark 4.3 *It is worth pointing out that in this chapter, $\frac{|z_i| z_i}{z_i^2 + \varepsilon(t)^2}$ is proposed to deal with the uncertain $y_r^{(j)}$ in the filter design. This is a smooth term playing a key role in coping with the time-varying trajectory in leader-following consensus, which guarantees asymptotic convergence of distributed errors z_i and does not belong to any existing robust compensating techniques.*

4.3 DESIGN OF LOCAL ADAPTIVE CONTROLLERS

The backstepping technique [56] will be applied for the design of the decentralized adaptive controller. Firstly introduce the error variables

$$e_{i,1} = x_{i,1} - q_{i,1},$$
$$e_{i,j} = x_{i,j} - \alpha_{i,j-1} - q_{i,j}, \quad j = 2, ..., n, \qquad (4.11)$$

where $\alpha_{i,j}$ are virtual controllers. The iterative controller design is described as follows:

Step 1. $\alpha_{i,1}$ is designed as

$$\alpha_{i,1} = -k_1 e_{i,1} - \varphi_{i,1}(x_{i,1})^T \hat{\theta}_i, \qquad (4.12)$$

where k_1 is a positive constant and $\hat{\theta}_i$ is the estimate of unknown parameter vector θ_i which will be designed later. Consider the Lyapunov function candidate

$$V_{i,1} = \frac{1}{2} e_{i,1}^2 + \frac{1}{2} \tilde{\theta}_i^T \Gamma_i^{-1} \tilde{\theta}_i, \qquad (4.13)$$

where Γ_i is a positive definite matrix and $\tilde{\theta}_i = \theta_i - \hat{\theta}_i$, then

$$\dot{V}_{i,1} = -k_1 e_{i,1}^2 + e_{i,1} e_{i,2} + \tilde{\theta}_i^T (\tau_{i1} - \Gamma_i^{-1} \dot{\hat{\theta}}_i), \qquad (4.14)$$

where $\tau_{i1} = e_{i,1} \varphi_{i,1}(x_{i,1})$ is a tuning function, as shown in [55].

Step j $(2 \le j \le n-1)$. Design the virtual control $\alpha_{i,j}$ as

$$\alpha_{i,j} = -k_j e_{i,j} - e_{i,j-1} + \sum_{l=1}^{j-1} \frac{\partial \alpha_{i,j-1}}{\partial x_{i,l}} x_{i,l+1} + \frac{\partial \alpha_{i,j-1}}{\partial \hat{\theta}_i} \Gamma_i \tau_{i,j}$$

$$+ \left(\sum_{l=2}^{j} \frac{\partial \alpha_{i,l-1}}{\partial \hat{\theta}_l} \right) \Gamma_i \left(\varphi_{i,j} - \sum_{l=1}^{j-1} \frac{\partial \alpha_{i,j-1}}{\partial x_{i,l}} \varphi_{i,l} \right)$$

$$- \hat{\theta}_i^T \left(\varphi_{i,j} - \sum_{l=1}^{j-1} \frac{\partial \alpha_{i,j-1}}{\partial x_{i,l}} \varphi_{i,l} \right) + \sum_{l=1}^{j-1} \frac{\partial \alpha_{i,j-1}}{\partial q_{i,l}} q_{i,l+1}, \tag{4.15}$$

where $\tau_{i,j} = \tau_{i,j-1} + \varphi_{i,j} - \sum_{l=1}^{j-1} \frac{\partial \alpha_{i,j-1}}{\partial x_{i,l}} \varphi_{i,l}$ is the tuning function. By defining the following Lyapunov function

$$V_{i,j} = V_{i,j-1} + \frac{1}{2} e_{i,j}^2, \tag{4.16}$$

then it has

$$\dot{V}_{i,j} = - \sum_{l=1}^{j} k_l e_{i,l}^2 + e_{i,j} e_{i,j+1} + \tilde{\theta}_i^T (\tau_{i,j} - \Gamma_i^{-1} \dot{\hat{\theta}}_i) \left(\sum_{l=2}^{j} e_{i,l} \frac{\partial \alpha_{i,l}}{\partial \hat{\theta}_i} \right) (\Gamma_i \tau_{i,j} - \dot{\hat{\theta}}_i). \tag{4.17}$$

Step n. The actual control u_i and parameter estimate $\hat{\theta}_i$ are designed as:

$$u_i = \alpha_{i,n} + v_i, \tag{4.18}$$

$$\dot{\hat{\theta}}_i = \Gamma_i \tau_{i,n},$$

$$\dot{\hat{D}}_i = \frac{e_{i,n}^2 |e_{i,n}|}{e_{i,n}^2 + \varepsilon^2}, \tag{4.19}$$

where $\hat{D}_i$ is the estimate of the external disturbance bound D_i, $\tilde{D}_i = D_i - \hat{D}_i$ and

$$\alpha_{i,n} = - k_j e_{i,n} - e_{i,n-1} + \sum_{l=1}^{n-1} \frac{\partial \alpha_{i,n-1}}{\partial x_{i,l}} x_{i,l+1}$$

$$+ \left(\sum_{l=2}^{n-1} \frac{\partial \alpha_{i,l-1}}{\partial \hat{\theta}_l} \right) \Gamma_i \left(\varphi_{i,n} - \sum_{l=1}^{n-1} \frac{\partial \alpha_{i,n-1}}{\partial x_{i,l}} \varphi_{i,l} \right)$$

$$- \hat{\theta}_i^T \left(\varphi_{i,n} - \sum_{l=1}^{n-1} \frac{\partial \alpha_{i,n-1}}{\partial x_{i,l}} \varphi_{i,l} \right) + \frac{\partial \alpha_{i,n-1}}{\partial \hat{\theta}_i} \Gamma_i \tau_{i,n}$$

$$+ \sum_{l=1}^{n-1} \frac{\partial \alpha_{i,n-1}}{\partial q_{i,l}} q_{i,l+1} - \frac{\hat{D}_i e_{i,n} |e_{i,n}|}{e_{i,n}^2 + \varepsilon^2}. \tag{4.20}$$

Consider the Lyapunov function

$$V_{i,n} = V_{i,n-1} + \frac{1}{2} e_{i,n}^2 + \frac{1}{2} \tilde{D}_i^2, \tag{4.21}$$

then it has

$$\dot{V}_{i,n} = - \sum_{l=1}^{n} k_l e_{i,l}^2 + \frac{1}{2} D_i \varepsilon(t). \tag{4.22}$$

Remark 4.4 *For simplicity and clarity, we only consider the matched disturbance case, i.e., the disturbance term $d_i(t)$ exists only in the last equation of (4.1). However, it is easy to extend the disturbances to be unmatched, i.e.,*

$$\dot{x}_{i,q} = x_{i,q+1} + \varphi_{i,q}(x_{i,1}, \ldots, x_{i,q})^T \theta_i + d_{i,q}(t), q = 1, \cdots, n-1,$$
$$\dot{x}_{i,n} = u_i + \varphi_{i,n}(x_i)^T \theta_i + d_{i,n}(t), \tag{4.23}$$

based on the smooth compensating term $\dfrac{|z_i| z_i^{n-q}}{z_i^{2(n-q+1)} + \varepsilon(t)^2}$ for qth equation of (4.23) in backstepping. It could be seen that $\dfrac{|z_i| z_i^{n-q}}{z_i^{2(n-q+1)} + \varepsilon(t)^2}$ is $(n-q)$th order derivable.

4.4 STABILITY AND CONSENSUS ANALYSIS

The main results of our distributed adaptive control design scheme can be formally stated in the following theorems.

Theorem 4.1 *Consider the closed-loop system consisting of N filters (4.2) satisfying Assumptions 4.1.1–4.1.2, with local controllers (4.8). All the signals in the closed-loop system are globally uniformly bounded and asymptotic consensus tracking of all the filters' outputs to $y_r(t)$ is achieved, i.e., $\lim_{t \to \infty}[q_{i,1}(t) - y_r(t)] = 0$.*

Proof. Consider the following Lyapunov function for the closed-loop system

$$V_1 = \frac{1}{2} z^T P z + \frac{1}{2\gamma} \sum_{j=1}^{n} \tilde{F}_j^T P \tilde{F}_j, \tag{4.24}$$

then

$$
\begin{aligned}
\dot{V}_1 = {}& z^T P(\mathcal{L} + \mathcal{B}) \Bigg(-cz - \sum_{j=1}^{n} \delta_j \mathrm{sgn}(z). * \underline{F}_j + \sum_{j=1}^{n} \delta_j \epsilon(z). * \underline{F}_j \\
& - \sum_{j=1}^{n} \delta_j Z. * \tilde{F}_j - \sum_{j=1}^{n} \delta_j \underline{y}_r^{(j)} \Bigg) - \frac{1}{\gamma} \sum_{j=1}^{n} \tilde{F}_j^T P(\mathcal{L} + \mathcal{B}) \tilde{F}_j \\
\leq {}& -cz^T Q z - \frac{1}{\gamma} \sum_{j=1}^{n} F_j^T Q \tilde{F}_j - \sum_{j=1}^{n} \delta_j z^T P \Delta \mathrm{sgn}(z). * \underline{F}_j \\
& + \sum_{j=1}^{n} \delta_j z^T P \mathcal{A} \underline{\mathrm{sgn}}(z). * \underline{F}_j - \sum_{j=1}^{n} \delta_j z^T P \mathcal{B} \underline{\mathrm{sgn}}(z). * \underline{F}_j \\
& - \sum_{j=1}^{n} \delta_j z^T P(\mathcal{L} + \mathcal{B}) \underline{y}_r^{(j)} - \sum_{j=1}^{n} \delta_j z^T P(\mathcal{L} + \mathcal{B}) \epsilon(z). * \underline{F}_j \\
& + \sum_{j=1}^{n} \delta_j \|z\| \| P(\mathcal{L} + \mathcal{B})\| \|\tilde{F}_j\|. \tag{4.25}
\end{aligned}
$$

It could be checked that

$$\sum_{j=1}^{n} \delta_j z^T P \Delta \underline{\text{sgn}(z)}. * \underline{F}_j = \sum_{j=1}^{n} \delta_j F_j \sum_{i=1}^{N} p_i \sum_{k=1}^{N} a_{ik} |\underline{z}_i|, \tag{4.26}$$

$$\sum_{j=1}^{n} \delta_j z^T P \mathcal{A} \underline{\text{sgn}(z)}. * \underline{F}_j \le \sum_{j=1}^{n} \delta_j F_j \sum_{i=1}^{N} p_i \sum_{k=1}^{N} a_{ik} |\underline{z}_i|, \tag{4.27}$$

$$\sum_{j=1}^{n} \delta_j z^T P \mathcal{B} \underline{\text{sgn}(z)}. * \underline{F}_j = \sum_{j=1}^{n} \delta_j F_j \sum_{i=1}^{N} p_i \mu_i |\underline{z}_i|, \tag{4.28}$$

$$\sum_{j=1}^{n} \delta_j z^T P(\mathcal{L} + \mathcal{B}) \underline{y}_r^{(j)} \le \sum_{j=1}^{n} \delta_j F_j \sum_{i=1}^{N} p_i \mu_i |\underline{z}_i|, \tag{4.29}$$

$$\|Z\| < 1. \tag{4.30}$$

Then,

$$\begin{aligned}
\dot{V}_1 \le & -cz^T Q z - \frac{1}{\gamma} \sum_{j=1}^{n} \tilde{F}_j^T Q \tilde{F}_j + \lambda_{\min}(Q) \sum_{j=1}^{n} \delta_j \|z\|^2 \\
& + \frac{\rho^2}{4\lambda_{\min}(Q)} \sum_{j=1}^{n} \delta_j \|\tilde{F}_j\|^2 - \sum_{j=1}^{n} \delta_j z^T P(\mathcal{L} + \mathcal{B}) \epsilon(z). * \underline{F}_j \\
\le & -\underline{c} \|z\|^2 - \bar{\gamma} \sum_{j=1}^{n} \|\tilde{F}_j\|^2 - \sum_{j=1}^{n} \delta_j z^T P(\mathcal{L} + \mathcal{B}) \epsilon(z). * \underline{F}_j, \tag{4.31}
\end{aligned}$$

where $\rho = \|P(\mathcal{L} + \mathcal{B})\|_2$, $\underline{c} = \lambda_{\min}(Q)(c - \sum_{j=1}^{n} \delta_j)$, and $\bar{\gamma} = \frac{1}{\gamma} - \frac{\rho^2}{4\lambda_{\min}(Q)}$ are positive constants, in which γ is a positive constant satisfying $\gamma < \frac{4\lambda_{\min}(Q)}{\rho^2}$.

Since $\|\epsilon(z)\| \le 1$, From (4.31) it is obvious that $\underline{z}_i$ and $\|\tilde{F}_j\|$ are bounded. Then

$$\dot{V}_1 \le -\bar{c} \|z\|^2 - \bar{\gamma} \sum_{j=1}^{n} \|\tilde{F}_j\|^2 + \varrho \|\epsilon(z)\|^2, \tag{4.32}$$

where $\bar{c} = \underline{c} - \frac{\lambda_{\min}(Q)}{2} \sum_{j=1}^{n} \delta_j = \lambda_{\min}(Q)(c - \frac{3}{2} \sum_{j=1}^{n} \delta_j)$ and $\varrho = \frac{\rho^2}{2\lambda_{\min}(Q)} \sum_{j=1}^{n} \delta_j F_j^2$. By choosing $c > \frac{3}{2} \sum_{j=1}^{n} \delta_j$, then $\bar{c} > 0$.

Based on the definition of $\epsilon(\underline{z}_i)$, it has $\epsilon(\underline{z}_i) = \text{sgn}(\underline{z}_i) - \frac{|\underline{z}_i| \underline{z}_i}{\underline{z}_i^2 + \varepsilon(t)^2} = \frac{\varepsilon(t)^2}{\underline{z}_i^2 + \varepsilon(t)^2} \text{sgn}(\underline{z}_i)$, then

$$\varrho \|\epsilon(z)\|^2 \le \frac{\varrho}{4} \varepsilon(t)^2, \tag{4.33}$$

thus it has

$$\dot{V}_1 \leq -\bar{c}\|z\|^2 - \bar{\gamma}\sum_{j=1}^n \|\tilde{F}_j\|^2 + \frac{\varrho\alpha^2}{4}e^{-2\beta t}, \qquad (4.34)$$

which means z, $\tilde{F}_j$, and V_1 are bounded. It can also be checked that $\dot{z}$ and $\dot{\tilde{F}}_j$ are also bounded. By taking integration of both sides of (4.34), it has

$$V_1(\infty) + \int_0^\infty \bar{c}\|z\|^2 d\tau + \int_0^\infty \bar{\gamma}\sum_{j=1}^n \|\tilde{F}_j\|^2 d\tau \leq V_1(0) + \frac{\varrho\alpha^2}{8\beta}, \qquad (4.35)$$

which means $\int_0^\infty \bar{c}\|z\|^2 d\tau$ and $\int_0^\infty \bar{\gamma}\sum_{j=1}^n \|\tilde{F}_j\|^2 d\tau$ are bounded. So, based on Barbalat's lemma, it has

$$\lim_{t\to\infty} \|z(t)\| = 0. \qquad (4.36)$$

Since $\|z_1\| = (\mathcal{L}+\mathcal{B})\|q_1 - \underline{y}_r\|$ and $\mathcal{L}+\mathcal{B}$ is nonsingular, then let $\sigma = \lambda_{\min}(\mathcal{L}+\mathcal{B})$, it has $\|q_1 - \underline{y}_r\| \leq \|z_1\|/|\sigma|$, thus

$$\lim_{t\to\infty}[q_{i,1}(t) - y_r(t)] = 0. \qquad (4.37)$$

Remark 4.5 *In [26, 64, 77], distributed consensus control problems of relatively simple multi-agent system models such as pure integrator dynamics and linear systems were considered. Signum functions of consensus errors were utilized to treat uncertain disturbances existing in system dynamics. However, the non-differentiability of these functions restricts such scheme being applicable to solve asymptotically consensus tracking problem for high-order nonlinear multi-agent systems in this chapter. To address this issue, a new smooth compensating term $\frac{z_i|z_i|}{z_i^2+\varepsilon(t)^2}$ associated with distributed consensus error variable z_i and a positive integrable time-varying function $\varepsilon(t)$, are introduced in designing v_i to account for the effects of unknown $y_r(t)$ for $\mu_i = 0$ and disturbances $d_i(t)$.*

Theorem 4.2 *Consider the closed-loop adaptive systems consisting of N uncertain nonlinear agents (4.1) satisfying Assumptions 4.1.1–4.1.2, the smooth controllers (4.18) and the parameter estimators (4.19). All the signals in the closed-loop system are globally uniformly bounded and asymptotic consensus tracking of all the agents' outputs to $y_r(t)$ is achieved, i.e., $\lim_{t\to\infty}[y_i(t) - y_r(t)] = 0$.*

Proof. From (4.22) and the definition of $V_{i,n}$, we can establish that $e_{i,l}$, $\hat{\theta}_i$ are bounded. Besides, $e_{i,l} \in \mathcal{L}_2$, thus $x_{i,1}$, $i = 1, ..., N$ are bounded. From (4.15) and (4.20), $\alpha_{i,j}$ and $\alpha_{i,n}$ are bounded, thus controller u_i is bounded. Therefore the boundedness of all the signals in the closed-loop adaptive systems is guaranteed. Thus $\dot{e}_{i,j}$, $j = 1, ..., n$ are bounded.

Integrating both sizes of (4.22) yields that

$$V_{i,n}(t) + \sum_{l=1}^{n} \int_0^t k_l e_{i,l}^2(\tau)d\tau = V_{i,n}(0) + \frac{D_i}{2}\int_0^t \alpha e^{-\beta\tau}d\tau, \qquad (4.38)$$

which means $\lim_{t\to\infty}\sum_{l=1}^{n}\int_0^t k_l e_{i,l}^2(\tau)d\tau$ is bounded. By applying Barbalat's lemma, we have $\lim_{t\to\infty} e_{i,j}(t) = 0$. Thus $\lim_{t\to\infty}[y_i(t) - q_{i,q}(t)] = 0$, therefore from (4.37)

$$\lim_{t\to\infty}[y_i(t) - y_r(t)] = 0. \qquad (4.39)$$

Remark 4.6 *From Theorem 4.2, it can be seen that global uniform boundedness of all the closed-loop system and asymptotically output consensus tracking can be achieved in this chapter. This is different from many existing results on consensus control of uncertain multi-agent systems including [30, 153, 161] where only semi-global uniform ultimate boundedness of tracking errors can be shown.*

4.5 SIMULATION RESULTS

We now consider an example to illustrate the proposed design schemes and verify the established theoretical results. Suppose that there are a group of four nonlinear agents with the following dynamics

$$\begin{aligned} \dot{x}_{i,1} &= x_{i,2} + \varphi_{i,1}(x_{i,1})\theta_i, \\ \dot{x}_{i,2} &= u_i + \varphi_{i,2}(x_{i,1}, x_{i,2})\theta_i + d_i(t), \qquad i = 1,\ldots,4 \end{aligned} \qquad (4.40)$$

where $\varphi_{i,1} = \sin(x_{i,1})$, $\varphi_{1,2} = x_{1,2}$, $\varphi_{2,2} = x_{2,2}^2$, $\varphi_{3,2} = x_{3,2}$, $\varphi_{4,2} = x_{4,1}x_{4,2}$. $\theta_1 = 1$, $\theta_2 = 0.5$, $\theta_3 = 2$, $\theta_4 = 3$, $d_1(t) = \sin(t)$, $d_2(t) = 2\sin(t)$, $d_3(t) = 2\sin^2(t)$ and $d_4(t) = 3\sin(t)$. The graph for these four agents is given in Figure 4.1.

The tracking performance of all the agents' outputs $y_i(t)$ for $1 \le i \le 4$ are shown in Figure 4.2. It can be seen that asymptotically consensus tracking can be achieved with the proposed distributed adaptive control scheme. Moreover, the boundedness of the the torques and parameter estimates can be observed in Figures 4.3–4.5, respectively. Figure 4.6 shows the comparison of torques with nonsmooth sgn($\cdot$)-type functions.

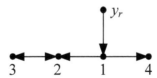

FIGURE 4.1 Gragh for a group of four nonlinear agents.

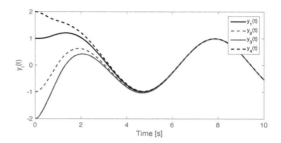

FIGURE 4.2 Tracking performance of the outputs $y_i(t)$ for $1 \leq i \leq 4$ to the desired trajectory $y_r(t)$.

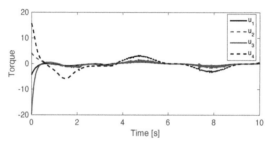

FIGURE 4.3 Torques of the four agents u_i for $1 \leq i \leq 4$.

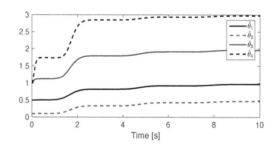

FIGURE 4.4 Parameter estimates $\hat{\theta}_i$ for $1 \leq i \leq 4$.

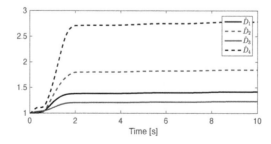

FIGURE 4.5 Disturbance estimates $\hat{D}_i$ for $1 \leq i \leq 4$.

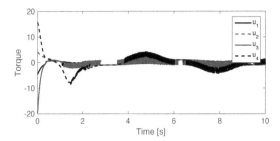

FIGURE 4.6 Comparison of torques of the four agents u_i with sgn($\cdot$)-type nonsmooth function [77].

4.6 NOTES

In this chapter, a new smooth distributed adaptive consensus tracking control scheme is proposed for nonlinear multi-agent systems with arbitrary system order, mismatched unknown parameters and uncertain external disturbances by introducing local estimators of the bounds of reference trajectory and a filter for each agent. Based on the scheme, asymptotically consensus output tracking can be achieved.

5 Hierarchical Decomposition-Based Distributed Adaptive Consensus Control

In Chapter 3, the distributed adaptive consensus tracking control problem for a class of uncertain high-order nonlinear multi-agent systems is considered and a distributed adaptive consensus tracking control scheme is proposed, with an assumption that the desired trajectory can be linearly parameterized with a set of basis functions which are known to all agents. This assumption is relaxed in Chapter 4 by introducing local estimators of the bounds of reference trajectory and a filter for each agent.

In this chapter, we investigate the consensus tracking control problem from a different perspective, i.e., we will make full use of the information contained within the graph and the assumptions mentioned in Chapter 3 can also be relaxed. Similar to Chapter 4, the desired trajectory $y_r(t)$ is known exactly by only part of the agents and the graph among the agents is directed. Compared with the results in Chapter 4 and some most related results in the literature, the main features of this chapter can be summarized as follows. First, the directed graph which represents the communication flow among the agents is split into a hierarchical structure according to the shortest possible path of each agent originated from the desired trajectory considered as a virtual leader node. By doing so, the controllers for the agents located in the upper and lower layers can be constructed in a sequential order with only locally available information utilized including that obtained from the neighbors. Moreover, in each agent, more estimators are designed to estimate the uncertainties of its neighbors' dynamics. The obtained estimators are then employed by the local adaptive controller of the agent. Thus the transmission of parameter estimates among connected agents is avoided. With the proposed design scheme, not only the boundedness of all closed-loop signals is ensured, but also asymptotically consensus tracking of all agent outputs is achieved.

5.1 PROBLEM FORMULATION

Similarly to [139] and Chapter 3, we consider a group of N nonlinear agents modeled as follows

$$\dot{x}_{i,q} = x_{i,q+1} + \varphi_{i,q}(x_{i,1}, \ldots, x_{i,q})^T \theta_i, \quad q = 1, \ldots, n-1$$
$$\dot{x}_{i,n} = b_i \beta_i(x_i) u_i + \varphi_{i,n}(x_{i,1}, \ldots, x_{i,n})^T \theta_i$$

DOI: 10.1201/9781003394372-5

$$y_i = x_{i,1}, \qquad \text{for } i = 1, \ldots, N \tag{5.1}$$

where $x_{i,q} \in \Re$ for $q = 1, \ldots, n$, $u_i \in \Re$ and $y_i \in \Re$ are the states, the control input and the output of the ith agent, respectively. $\theta_i \in \Re^{p_i}$ is the vector of unknown constants and the high frequency gain $b_i \neq 0 \in \Re$ is an unknown constant. $\varphi_{i,q} : \Re^i \to \Re^{p_i}$, for $q = 1, \ldots, n$ and $\beta_i : \Re^n \to \Re^1$ are known smooth nonlinear functions.

The desired trajectory for the outputs of the group of N agents is given by $y_r(t)$ whose first nth derivatives are assumed bounded and only available to a small percentage of the agents in $\mathcal{G}$. The *control objectives* are to design distributed adaptive controllers u_i for all N agents (i.e., $1 \leq i \leq N$) by using only locally available information including that obtained from the neighbors such that:

- The outputs of all N agents in $\mathcal{G}$ can reach a consensus by tracking the desired trajectory $y_r(t)$ asymptotically;
- The closed-loop system is globally stable in the sense that all the signals in the closed loop are uniformly ultimately bounded.

To achieve the objectives, the following assumptions are imposed.

Assumption 5.1.1 *Let $\bar{\mathcal{G}} = (\bar{\mathcal{V}}, \bar{\mathcal{E}})$ be the augmented directed graph consisting of $\mathcal{G}$ and $y_r(t)$ considered as a virtual leader node 0, i.e., $\bar{\mathcal{V}} = \{0, 1, \ldots, N\}$. Then $y_r(t)$ has directed paths to all the agents in $\mathcal{G}$.*

Assumption 5.1.2 *The sign of b_i for $i = 1, \ldots, N$ are known.*

Remark 5.1 *From Assumption 5.1.1, there is at least one agent in $\mathcal{G}$ that can capture the information $y_r(t)$ directly. Combined with the definitions in [112], it indicates that $\bar{\mathcal{G}}$ has a directed spanning tree in which $y_r(t)$ is the root with only outgoing edges. Note that Assumption 5.1.1 is a weaker condition than the original communication graph $\mathcal{G}$ itself having a directed spanning tree [139] or being strongly connected [16].*

5.2 HIERARCHICAL DESIGN OF DISTRIBUTED ADAPTIVE CONTROLLERS

5.2.1 HIERARCHICAL DECOMPOSITION

As a preliminary work to the design of distributed adaptive controllers, we will first split the directed graph $\mathcal{G}$ into a hierarchical structure with r layers, where $1 \leq r \leq N$, according to the shortest possible path corresponding to each agent (also known as node) in the augmented graph $\bar{\mathcal{V}}$ originated from the virtual leader node $y_r(t)$. The detailed procedure is given as follows.

⋄ *Hierarchical Decomposition Algorithm:*

1. Initialization: the virtual leader node $y_r(t)$ is classified as "Layer 0";
2. For an agent in $\mathcal{G}$, if it has a direct access to $y_r(t)$, it is classified as "Layer 1" and labeled as "l_1".

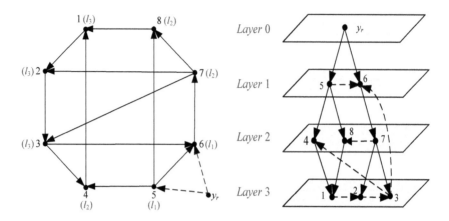

FIGURE 5.1 Original graph for a group of 8 agents.

FIGURE 5.2 The hierarchical structure obtained from Figure 5.1, in which the maintained edges from the original graph are marked with dark black lines.

3. For the rest agents in $\mathcal{G}$, if it can obtain information from the agents labeled as l_{q-1}, for $q = 2, \ldots, r$, it is classified as "Layer q" labeled as "l_q".
4. Once the agent is labeled, it focuses on the information obtained from the agents in the upper layer only and the algorithm ends locally.

By the hierarchical decomposition algorithm above, a new graph $\mathcal{G}_n$ for all the N agents is formed by maintaining only the edges in $\mathcal{E}$ originated from the upper layers to the first lower layers. Similarly, $\bar{\mathcal{G}}_n$ represents the new augmented graph by including the virtual leader node $y_r(t)$. We now use an example to illustrate our idea of hierarchical decomposition.

Example 5.1 *Suppose that the original graph for a group of 8 agents is shown in Figure 5.1. $y_r(t)$ is available to agent 5 and 6. After hierarchical decomposition presented above, the 8 agents will be classified into three subgroups, i.e., Layers 1 to 3 as shown in Figure 5.2. It can been seen that only the edges between the upper layers and first lower layers are maintained in constructing the new directed graph. Moreover, hierarchical decomposition will not change the property that $y_r(t)$ has directed paths to all of the agents.*

Define a vector $\mu = [\mu_1, \mu_2, \ldots, \mu_N]^T \in \Re^N$, where $\mu_i > 0$ if $y_r(t)$ is available to agent i in $\mathcal{G}$ and $\mu_i = 0$ otherwise. Let the Laplacian matrices of the new directed graphs $\mathcal{G}_n$ and $\bar{\mathcal{G}}_n$ be $\mathcal{L}_n \in \Re^{N \times N}$ and $\bar{\mathcal{L}}_n \in \Re^{(N+1) \times (N+1)}$, respectively. We then have

$$\bar{\mathcal{L}}_n = \begin{bmatrix} 0 & 0_{1 \times N} \\ -\mu & \mathcal{L}_n + \mathcal{B} \end{bmatrix} \tag{5.2}$$

where $\mathcal{B} = \text{diag}\{\mu_i\}$. By following similar analysis in [112], we have the following result.

Lemma 5.1 *Consider a group of agents with the desired trajectory $y_r(t)$ under Assumption 5.1.1. By the hierarchical decomposition algorithm 1) − 4), the obtained Laplacian matrix $\mathcal{L}_n$ satisfies that $(\mathcal{L}_n + \mathcal{B})$ has full rank and all eigenvalues of $(\mathcal{L}_n + \mathcal{B})$ have positive real parts.*

5.2.2 DESIGN OF DISTRIBUTED ADAPTIVE CONTROLLERS

Suppose that the sets of agents in Layer j are denoted by P_j for $j = 1, \ldots, r$ where r is the number of layers after performing hierarchical decomposition in previous section. It is obvious that $\sum_{j=1}^{r} n(P_j) = N$, where $n(P_j)$ is the number of agents in Layer j.

⋄ For agent i in Layer 1, i.e., $i \in P_1$, the distributed controller design is achieved by following similar procedure in [56] and are summarized in Table 5.1. $c_{i,q}$, $q = 1, \ldots, n$, γ_i are positive constants and Γ_i is a positive definite matrix. $\hat{\varrho}_i$ and $\hat{\theta}_i$ are the estimates of $\varrho_i = 1/b_i$ and θ_i, respectively.

⋄ For agent i in Layer $j(\geq 2)$, i.e., $i \in P_j$, $j = 2, \ldots, r$, the output consensus error is defined in the first step by using local available information obtained from the agents k in Layer $j - 1$, i.e., $z_{i,1} = f(x_{i,1}, a_{ik}x_{k,1})$ for $k \in P_{j-1}$. Moreover, extra local estimates $\hat{b}_{ik}$, $\hat{\theta}_{ik}$ for the unknown external parameters b_k, θ_k in agent k for $k \in P_{j-1}$ are introduced in designing the distributed adaptive controller u_i if $a_{ik} = 1$. The detailed design of u_i for $i \in P_j$ are summarized in Table 5.2. Similar to Table 5.1, $c_{i,q}$, $q = 1, \ldots, n$, γ_i, γ_{ik} for $k \in P_{j-1}$ are positive constants and Γ_i, Γ_{ik} for $k \in P_{j-1}$ are positive definite matrix. $\hat{\varrho}_i$ and $\hat{\theta}_i$ are the estimates of $\varrho_i = 1/b_i$, θ_i, respectively.

Remark 5.2 *Note that in Table 5.2, the fact that $\alpha_{i,q}$ for $q = 1, \ldots, n - 1$, $i \in P_j$, $j = 2, \ldots, r$ is a function of $\bar{x}_{i,q} = (x_{i,1}, \ldots, x_{i,q})$, $a_{ik}\bar{x}_{k,q+1} = (x_{k,1}, \ldots, x_{k,q+2})$, $\hat{\theta}_i$ and $a_{ik}\hat{\theta}_{ik}$ for $k \in P_{j-1}$ was used. Similar to [139], the structure information of the neighbors' intrinsic dynamics such as $\varphi_{k,q}$ need to be collected if $a_{ik} = 1$. However, by introducing the estimates in an agent for the uncertainties of its neighbors' dynamics, the information exchange of local parameter estimates is not required, in contrast to [139]. Furthermore, it can be seen from (5.20) and (5.29) that in constructing the adaptive controllers for agent $i \in P_j$, the control inputs generated by its neighbors located in the upper layer, i.e., u_k if $a_{ik} = 1$ for $k \in P_{j-1}$, is also utilized. This implies that by hierarchical decomposition, the local controllers for the agents in the upper and lower layers can be generated in a sequential order with only locally available information adopted including that obtained from the neighbors.*

TABLE 5.1

Adaptive Backstepping Control Design for the Agents in Layer 1, i.e., $i \in P_1$

Introducing Error Variables:

$$z_{i,1} = y_i - y_r \tag{5.3}$$

$$z_{i,q} = x_{i,q} - \alpha_{i,q-1} - y_r^{(q-1)}, \quad q = 2, 3, \ldots, n \tag{5.4}$$

Control Laws:

$$u_i = \frac{\hat{\varrho}_i}{\beta_i}\left(\alpha_{i,n} + y_r^{(n)}\right) \tag{5.5}$$

$$\alpha_{i,1} = -c_{i,1}z_{i,1} - \varphi_{i,1}^T\hat{\theta}_i \tag{5.6}$$

$$\alpha_{i,q} = -z_{i,q-1} - c_{i,q}z_{i,q} - \delta_{i,q}^T\hat{\theta}_i + \frac{\partial \alpha_{i,q-1}}{\partial \hat{\theta}_i}\Gamma_i\tau_{i,q}$$

$$+ \sum_{l=1}^{q-1}\left(\frac{\partial \alpha_{i,q-1}}{\partial x_{i,l}}x_{i,l+1} + \frac{\partial \alpha_{i,q-1}}{\partial y_r^{(l-1)}}y_r^{(l)}\right)$$

$$+ \sum_{l=2}^{q-1}\frac{\partial \alpha_{i,l-1}}{\partial \hat{\theta}_i}\Gamma_i\delta_{i,q}z_{i,l}, \quad q = 2, 3, \ldots, n \tag{5.7}$$

$$\delta_{i,q} = \varphi_{i,q} - \sum_{l=1}^{q-1}\frac{\partial \alpha_{i,q-1}}{\partial x_{i,l}}\varphi_{i,l}, \quad q = 2, 3, \ldots, n \tag{5.8}$$

$$\tau_{i,1} = \varphi_{i,1}z_{i,1} \tag{5.9}$$

$$\tau_{i,q} = \tau_{i,q-1} + \delta_{i,q}z_{i,q}, \quad q = 2, 3, \ldots, n \tag{5.10}$$

Parameter Update Laws:

$$\dot{\hat{\varrho}}_i = -\gamma_i\text{sgn}(b_i)\left(\alpha_{i,n} + y_r^{(n)}\right)z_{i,n} \tag{5.11}$$

$$\dot{\hat{\theta}}_i = \Gamma_i\tau_{i,n} \tag{5.12}$$

TABLE 5.2

Adaptive Backstepping Control Design for the Agents in Layer $j(\geq 2)$, **i.e.,** $i \in P_j$ **for** $j = 2, \ldots, r$

Introducing Error Variables:

$$z_{i,1} = \sum_{k \in P_{j-1}} a_{ik}(x_{i,1} - x_{k,1}) \tag{5.13}$$

$$z_{i,q} = x_{i,q} - \alpha_{i,q-1}, \quad q = 2, 3, \ldots, n \tag{5.14}$$

Control Laws:

$$u_i = \frac{\hat{\varrho}_i}{\beta_i} \alpha_{i,n} \tag{5.15}$$

$$\alpha_{i,1} = \frac{\bar{\alpha}_{i,1}}{\bar{d}_i} - \varphi_{i,1}^T \hat{\theta}_i, \quad \bar{d}_i = \sum_{k \in P_{j-1}} a_{ik} \tag{5.16}$$

$$\bar{\alpha}_{i,1} = -c_{i,1}z_{i,1} + \sum_{k \in P_{j-1}} a_{ik}\left(x_{k,2} + \varphi_{k,1}^T \hat{\theta}_{ik}\right) \tag{5.17}$$

$$\alpha_{i,2} = -\bar{d}_i z_{i,1} - c_{i,2} z_{i,2} - \delta_{i,2}^T \hat{\theta}_i + \frac{\partial \alpha_{i,1}}{\partial \hat{\theta}_i} \Gamma_i \tau_{i,2}$$

$$+ \frac{\partial \alpha_{i,1}}{\partial x_{i,1}} x_{i,2} + \sum_{k \in P_{j-1}} a_{ik} \left[\sum_{l=1}^{2} \frac{\partial \alpha_{i,1}}{\partial x_{k,l}} x_{k,l+1} \right.$$

$$\left. + \delta_{ik,2}^T \hat{\theta}_{ik} + \frac{\partial \alpha_{i,1}}{\partial \hat{\theta}_{ik}} \Gamma_{ik} \tau_{ik,2} \right]$$
$$\tag{5.18}$$

$$\alpha_{i,q} = -z_{i,q-1} - c_{i,q} z_{i,q} - \delta_{i,q}^T \hat{\theta}_i + \frac{\partial \alpha_{i,q-1}}{\partial \hat{\theta}_i} \Gamma_i \tau_{i,q}$$

$$+ \sum_{l=1}^{q-1} \frac{\partial \alpha_{i,q-1}}{\partial x_{i,l}} x_{l+1} + \sum_{l=2}^{q-1} \frac{\partial \alpha_{i,l-1}}{\partial \hat{\theta}_i} \Gamma_i \delta_{i,q} z_{i,l}$$

$$+ \sum_{k \in P_{j-1}} a_{ik} \left[\sum_{l=1}^{q} \frac{\partial \alpha_{i,q-1}}{\partial x_{k,l}} x_{k,l+1} + \delta_{ik,q}^T \hat{\theta}_{ik} \right.$$

$$\left. + \frac{\partial \alpha_{i,q-1}}{\partial \hat{\theta}_{ik}} \Gamma_{ik} \tau_{ik,q} + \sum_{l=2}^{q-1} \frac{\partial \alpha_{i,l-1}}{\partial \hat{\theta}_{ik}} \Gamma_{ik} \delta_{ik,q} z_{i,l} \right],$$

$$q = 3, \ldots, n-1 \tag{5.19}$$

(Continued)

TABLE 5.2

Adaptive Backstepping Control Design for the Agents in Layer $j (\geq 2)$, **i.e.,** $i \in P_j$ **for** $j = 2, \ldots, r$ **(Continued)**

$$
\begin{aligned}
\alpha_{i,n} = {}& -z_{i,n-1} - c_{i,n} z_{i,n} - \delta_{i,n}^T \hat{\theta}_i + \frac{\partial \alpha_{i,n-1}}{\partial \hat{\theta}_i} \Gamma_i \tau_{i,n} \\
& + \sum_{l=1}^{n-1} \frac{\partial \alpha_{i,n-1}}{\partial x_{i,l}} x_{l+1} + \sum_{l=2}^{n-1} \frac{\partial \alpha_{i,l-1}}{\partial \hat{\theta}_i} \Gamma_i \delta_{i,n} z_{i,l} \\
& + \sum_{k \in P_{j-1}} a_{ik} \left[\sum_{l=1}^{n-1} \frac{\partial \alpha_{i,n-1}}{\partial x_{k,l}} x_{k,l+1} + \frac{\partial \alpha_{i,n-1}}{\partial x_{k,n}} \hat{b}_{ik} \right. \\
& \qquad\qquad \times \beta_k u_k + \delta_{ik,n}^T \hat{\theta}_{ik} + \frac{\partial \alpha_{i,n-1}}{\partial \hat{\theta}_{ik}} \Gamma_{ik} \tau_{ik,n} \\
& \qquad\qquad \left. + \sum_{l=2}^{n-1} \frac{\partial \alpha_{i,l-1}}{\partial \hat{\theta}_{ik}} \Gamma_{ik} \delta_{ik,n} z_{i,l} \right] \qquad\qquad (5.20)
\end{aligned}
$$

$$
\delta_{i,q} = \varphi_{i,q} - \sum_{l=1}^{q-1} \frac{\partial \alpha_{i,q-1}}{\partial x_{i,l}} \varphi_{i,l}, \quad q = 2, 3, \ldots, n \qquad (5.21)
$$

$$
\delta_{ik,q} = \sum_{l=1}^{q} \frac{\partial \alpha_{i,q-1}}{\partial x_{k,l}} \varphi_{k,l}, \quad q = 2, 3, \ldots, n \qquad (5.22)
$$

$$
\tau_{i,1} = \varphi_{i,1} z_{i,1} \qquad\qquad (5.23)
$$

$$
\tau_{ik,1} = -\varphi_{k,1} z_{i,1} \qquad\qquad (5.24)
$$

$$
\tau_{i,q} = \tau_{i,q-1} + \frac{\delta_{i,q}}{\bar{d}_i} z_{i,q}, \quad q = 2, 3, \ldots, n \qquad (5.25)
$$

$$
\tau_{ik,q} = \tau_{ik,q-1} - \delta_{ik,q} z_{i,q}, \quad q = 2, 3, \ldots, n \qquad (5.26)
$$

Parameter Update Laws:

$$
\dot{\hat{\varrho}}_i = -\gamma_i \mathrm{sgn}(b_i) \alpha_{i,n} z_{i,n} \qquad\qquad (5.27)
$$

$$
\dot{\hat{\theta}}_i = \Gamma_i \tau_{i,n} \qquad\qquad (5.28)
$$

$$
\dot{\hat{b}}_{ik} = -\gamma_{ik} \frac{\partial \alpha_{i,n-1}}{\partial x_{k,n}} \beta_k u_k z_{i,n}, \quad k \in P_{j-1} \qquad (5.29)
$$

$$
\dot{\hat{\theta}}_{ik} = \Gamma_{ik} \tau_{ik,n}, \quad k \in P_{j-1} \qquad\qquad (5.30)
$$

5.3 STABILITY AND CONSENSUS ANALYSIS

It will be shown in this section that in contrast to the results in [139], the output consensus tracking will be achieved asymptotically in addition to the boundedness of all closed-loop signals.

◇ For agent i in Layer 1, i.e., $i \in P_1$, we choose the Lyapunov function as

$$V_i = \sum_{q=1}^{n} \frac{1}{2} z_{i,q}^2 + \frac{1}{2} \tilde{\theta}_i^T \Gamma_i^{-1} \tilde{\theta}_i + \frac{b_i}{2\gamma_i} \tilde{\varrho}_i^2 \tag{5.31}$$

where $\tilde{\theta}_i = \theta_i - \hat{\theta}_i$ and $\tilde{\varrho}_i = \varrho_i - \hat{\varrho}_i$ are estimation errors. Then from (5.3)–(5.12), the derivative of V_i is given by

$$\dot{V}_i = -\sum_{q=1}^{n} c_{i,q} z_{i,q}^2 + \left(\sum_{l=2}^{n} z_{i,l} \frac{\partial \alpha_{i,l-1}}{\partial \hat{\theta}_i} \right) \left(\Gamma_i \tau_{i,n} - \dot{\hat{\theta}}_i \right)$$

$$+ \tilde{\theta}_i^T \Gamma_i^{-1} \left(\Gamma_i \tau_{i,n} - \dot{\hat{\theta}}_i \right) - \frac{|b_i|}{\gamma_i} \tilde{\varrho}_i \left(\gamma_i \mathrm{sgn}(b_i) \alpha_{i,n} z_{i,n} + \dot{\hat{\varrho}}_i \right)$$

$$= -\sum_{q=1}^{n} c_{i,q} z_{i,q}^2 \leq 0. \tag{5.32}$$

From the definition of V_i for $i \in P_1$ in (5.31), we can establish that $z_{i,q}$ for $q = 1, \ldots, n$, $\hat{\theta}_i$, $\hat{\varrho}_i$ are bounded. From (5.3) and the boundedness of y_r, it follows that $x_{i,1}$ is bounded. From (5.6) and the smoothness of φ_i, $\alpha_{i,1}$ is also bounded. From the fact that $x_{i,2} = z_{i,2} + \alpha_{i,1} + \dot{y}_r$, $x_{i,2}$ is bounded. By following similar procedure, the boundedness of $\alpha_{i,q}$ for $q = 2, \ldots, n$, $x_{i,q}$ for $q = 3, \ldots, n$ is ensured. From (5.5) and the smoothness of β_i, we can conclude that the control signal u_i is bounded.

◇ For agent i in Layer $j(\geq 2)$, i.e., $i \in P_j$ for $j = 2, \ldots, r$, we choose the Lyapunov function as

$$V_i = \sum_{q=1}^{n} \frac{1}{2} z_{i,q}^2 + \frac{\bar{d}_i}{2} \tilde{\theta}_i^T \Gamma_i^{-1} \tilde{\theta}_i + \sum_{k \in P_{j-1}} a_{ik} \tilde{\theta}_{ik}^T \Gamma_{ik}^{-1} \tilde{\theta}_{ik}$$

$$+ \sum_{k \in P_{j-1}} a_{ik} \frac{1}{2\gamma_{ik}} \tilde{b}_{ik}^2 + \frac{b_i}{2\gamma_i} \tilde{\varrho}_i^2 \tag{5.33}$$

where $\tilde{\theta}_i$, $\tilde{\varrho}_i$ are estimation errors defined similarly to the case when $i \in P_1$. $\tilde{b}_{ik} = b_{ik} - \hat{b}_{ik}$, $\tilde{\theta}_{ik} = \theta_k - \hat{\theta}_{ik}$, $k \in P_{j-1}$ if $a_{ik} = 1$. Then from (5.13)–(5.30), the derivative of v_i is given by

$$\dot{V}_i = -\sum_{q=1}^{n} c_{i,q} z_{i,q}^2 + \left(\sum_{l=2}^{n} z_{i,l} \frac{\partial \alpha_{i,l-1}}{\partial \hat{\theta}_i} \right) \left(\Gamma_i \tau_{i,n} - \dot{\hat{\theta}}_i \right) + \bar{d}_i \tilde{\theta}_i^T \Gamma_i^{-1} \left(\Gamma_i \tau_{i,n} - \dot{\hat{\theta}}_i \right)$$

$$+ \sum_{k \in P_{j-1}} a_{ik} \left(\sum_{l=2}^{n} z_{i,l} \frac{\partial \alpha_{i,l-1}}{\partial \hat{\theta}_{ik}} \right) \left(\Gamma_{ik} \tau_{ik,n} - \dot{\hat{\theta}}_{ik} \right)$$

$$+ \sum_{k \in P_{j-1}} a_{ik} \tilde{\theta}_{ik}^T \Gamma_{ik}^{-1} \left(\Gamma_{ik} \tau_{ik,n} - \dot{\hat{\theta}}_{ik} \right) - \sum_{k \in P_{j-1}} a_{ik} \frac{\tilde{b}_{ik}}{\gamma_{ik}} \left(\gamma_{ik} \frac{\partial \alpha_{i,n-1}}{x_{k,n}} \beta_k u_k \right.$$

$$\left. + \dot{\hat{b}}_{ik} \right) - \frac{|b_i|}{\gamma_i} \tilde{\varrho}_i \left(\gamma_i \text{sgn}(b_i) \alpha_{i,n} z_{i,n} + \dot{\hat{\varrho}}_i \right)$$

$$= - \sum_{q=1}^{n} c_{i,q} z_{i,q}^2 \leq 0. \tag{5.34}$$

From the definition of V_i for $i \in P_2$ in (5.33), we can establish that $z_{i,q}$ for $q = 1, \ldots, n$, $\hat{\theta}_i$, $\hat{\varrho}_i$ and $\hat{b}_{ik}$, $\hat{\theta}_{ik}$ for $k \in P_1$ are bounded. From (5.13) and the boundedness of $x_{k,1}$ for $k \in P_1$, it follows that $x_{i,1}$ for $i \in P_2$ is bounded. From (5.17) and smoothness of φ_k, $\bar{\alpha}_{i,1}$ is bounded. Combining with the smoothness of φ_i, $\alpha_{i,1}$ for $i \in P_2$ is also bounded. Similarly to the agents in Layer 1, the boundedness of $\alpha_{i,q}$, $x_{i,q}$, $q = 2, \ldots, n$ as well as the control inputs u_i for $i \in P_2$ is ensured. By repeating the analyzing procedure above in a recursive manner, the boundedness of the signals in the rest agents i for $i \in P_j$, $j = 3, \ldots, r$ can be shown. Designing the Lyapunov function for the whole group of agents as $V = \sum_{i=1}^{N} V_i$, we have that $\dot{V} = -\sum_{i=1}^{N} \sum_{q=1}^{n} c_{i,q} z_{i,q}^2$. By applying the LaSalle-Yoshizawa theorem, it further follows that $\lim_{t \to \infty} z_{i,q}(t) = 0$ for $i = 1, \ldots, N$ and $q = 1, \ldots, n$. Observing the definitions of $z_{i,1}$ in (5.3) for $i \in P_1$ and in (5.13) for $i \in P_j$, $j = 2, \ldots, r$, it follows that

$$z_1 = (\mathcal{L}_n + \mathcal{B})(y - \underline{y}_r), \tag{5.35}$$

where $z_1 = [z_{1,1}, \ldots, z_{N,1}]^T$, $y = [y_1, \ldots, y_N]^T$ and $\underline{y}_r = y_r \times [1, \ldots, 1]^T = [y_r, \ldots, y_r]^T$. From Lemma 5.1, we have that $y - \underline{y}_r = (\mathcal{L}_n + \mathcal{B})^{-1} z_1$ and $\lim_{t \to \infty} [y_i(t) - y_r(t)] = 0$ for $i = 1, \ldots, N$. This implies that asymptotic consensus tracking of all agent outputs to a desired trajectory y_r is achieved. The results above are summarized in the following theorem.

Theorem 5.1 *Consider a group of N uncertain nonlinear agents as modeled in (5.1) with a desired trajectory $y_r(t)$ under Assumption 5.1.1–5.1.2. By performing the hierarchical decomposition algorithm 1) − 4), designing the distributed adaptive controllers in (5.5), (5.15), the parameter update laws in (5.11)–(5.12), (5.27)– (5.30), all closed-loop signals are globally uniformly bounded and the asymptotic consensus tracking of all agent outputs to $y_r(t)$ is achieved, i.e., $\lim_{t \to \infty} [y_i(t) - y_r(t)] = 0$ for $i = 1, \ldots, N$.*

5.4   SIMULATION RESULTS

We consider a group of five agents with the following intrinsic dynamics

$$\dot{x}_{i,1} = x_{i,2} + \sin(x_{i,1})\theta_i$$
$$\dot{x}_{i,2} = x_{i,2}^2 + b_i u_i \tag{5.36}$$

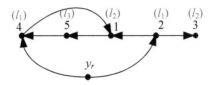

FIGURE 5.3 Original graph for a group of five agents [139].

where $\theta_1 = 1$, $\theta_2 = 1.5$, $\theta_3 = -2$, $\theta_4 = -3$, $\theta_5 = -1$, $b_1 = 1$, $b_2 = -2$, $b_3 = 3$, $b_4 = 2.5$, and $b_5 = -4$ are unknown system parameters. The communication graph corresponding to the network is shown in Figure 5.3, in which the layer number for each agent by hierarchical decomposition is indicated. The desired trajectory is $y_r = 0.3\sin(0.1t)$. The initials of all agent outputs are set as $[y_1(0), y_2(0), y_3(0), y_4(0), y_5(0)]^T = [0.7, 0.4, 0, -0.2, -0.6]^T$. The design parameters are chosen as $c_{i,1} = 8$, $c_{i,2} = 3$, $\gamma_i = \Gamma_i = \Gamma_{ik} = \gamma_{ik} = 0.05$ if $a_{ik} = 1$ for $i, k = 1, \ldots, 5$. The responses of all agent outputs $y_i(t)$ for $1 \le i \le 5$ with comparison to the desired trajectory $y_r(t)$ are shown in Figure 5.4. The corresponding performances of the agent inputs $u_i(t)$, the states $x_{i,2}(t)$ and the parameter estimates for all agents are given in Figures 5.5, 5.6 and 5.7, respectively. It can be seen that perfect consensus tracking is achieved while all the signals are bounded with the proposed method.

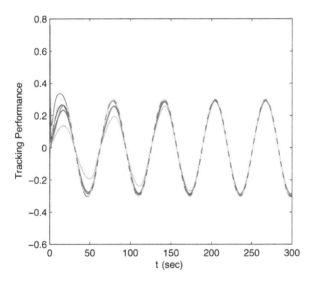

FIGURE 5.4 Tracking performance of the outputs $y_i(t)$ (solid lines) for $1 \le i \le 5$ to the desired trajectory $y_r(t)$ (dashed line).

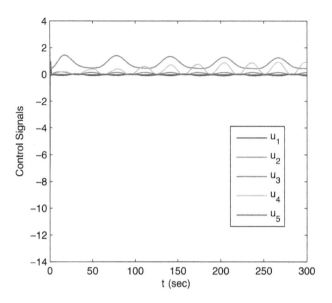

FIGURE 5.5 Input signals $u_i(t)$ for $1 \le i \le 5$.

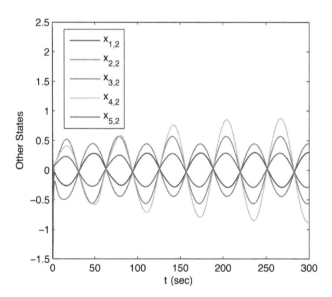

FIGURE 5.6 The states $x_{i,2}(t)$ for $1 \le i \le 5$.

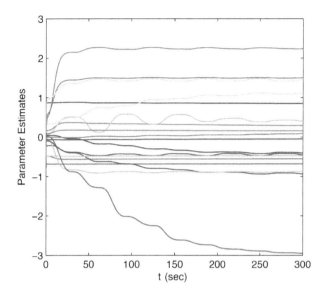

FIGURE 5.7 Parameter estimates in all agents including $\hat{\theta}_i$, $\hat{\varrho}_i$ for $1 \leq i \leq 5$, $\hat{\theta}_{12}$, $\hat{\theta}_{14}$, $\hat{\theta}_{32}$, $\hat{\theta}_{51}$, $\hat{b}_{12}$, $\hat{b}_{14}$, $\hat{b}_{32}$, and $\hat{b}_{51}$.

5.5 NOTES

In this chapter, a hierarchical decomposition-based distributed adaptive control scheme is proposed for a team of uncertain nonlinear agents. The local controllers for the agents in upper and lower layers can be constructed in a sequential order. By estimating the unknown parameters of the neighbors' intrinsic dynamics in an agent for designing its local adaptive controller, the transmission of local parameter estimates among connected agents is avoided. It is proved that boundedness of all closed-loop signals is ensured and asymptotical output consensus tracking can be achieved with our proposed scheme.

Section II

Finite-Time Consensus Control

6 Adaptive Finite-Time Consensus for Uncertain Nonlinear Mechanical Systems

In the previous chapters including Chapters 3, 4, and 5, we have investigated the distributed adaptive consensus tracking control problem for a class of high-order nonlinear multi-agent systems with unmatched unknown parameters. However, the proposed control schemes can only guarantee the asymptotic convergence of the consensus errors. This implies that the convergence rate at best is exponential and the consensus errors need infinite time to converge to zero.

In practice, it is often required that the consensus errors converge to zero in finite time. This motivates the investigation of finite-time consensus control. As discussed in Chapter 1, a number of researchers have investigated the finite-time consensus control problems; see [10, 60, 129, 143, 166] for instance. However, it should be pointed out that most of the existing results did not consider uncertainties including unmodeled system dynamics and unknown system parameters. Although adaptive control has been proven to be an effective tool to cope with uncertainties, achieving adaptive finite-time control is notably challenging. Certain key techniques employed in existing adaptive control cannot be applied. For example, the Barbalat's lemma normally adopted for analyzing asymptotic convergence cannot be applied to the analysis of finite-time convergence. On the other hand, the finite-time convergence analysis tools adopted in systems without parametric uncertainties such as those in [9] and [60] cannot be applied to adaptive finite-time control directly.

In this chapter, we consider finite-time leaderless consensus control problem for a group of general nonlinear mechanical systems with parametric uncertainties. A novel continuous adaptive distributed finite-time controllers are proposed for each agent in leaderless consensus control. Besides, suitable online parameter estimators are designed to guarantee that the position errors and the virtual control errors converge to a pre-defined compact set within finite time. It is shown that the states of the mechanical systems can reach a consensus within finite time under an undirected graph. Transient performances in terms of convergence rates and time are also analyzed. Finally simulation results illustrate and verify the effectiveness of the proposed schemes.

DOI: 10.1201/9781003394372-6

6.1 PRELIMINARIES

To facilitate the analysis of finite-time stability in the sequel, some basic concepts and definitions are firstly introduced.

Definition 6.1 ([9]) *Consider a dynamic system*

$$\dot{x} = f(x,t), \ f(0,t) = 0, \ x \in U_0 \subset R^n, \tag{6.1}$$

where $f : U_0 \times R^+ \to R^n$ is continuous on an open neighborhood U_0 of the origin $x = 0$. The equilibrium $x = 0$ of the system is (locally) finite-time stable if it is Lyapunov stable and for any initial condition $x(t_0) = x_0 \in U$ where $U \subset U_0$, if there is a settling time $T > t_0$, such that every solution $x(t; t_0, x_0)$ of system (6.1) satisfies $x(t; t_0, x_0) \in U \backslash \{0\}$ for $t \in [t_0, T)$, and

$$\lim_{t \to T} x(t; t_0, x_0) = 0, \ x(t; t_0, x_0) = 0, \ \forall t > T.$$

If $U = R^n$, then the origin $x = 0$ is a globally finite-time stable equilibrium.
The following lemmas are useful for establishing system stability.

Lemma 6.1 ([9]) *Suppose there is a C^1 positive definite Lyapunov function $V(x,t)$ defined on $U \times R^+$ where $U \subset U_0$ is the neighborhood of the origin, and there are positive real constants $c > 0$ and $0 < \alpha < 1$, such that $\dot{V}(x,t) + cV^\alpha(x,t)$ is negative semi-definite on U. Then $V(x,t)$ is locally in finite-time convergent with a settling time*

$$T \le \frac{V^{1-\alpha}(x_0, t)}{c(1 - \alpha)},$$

for any given initial condition $x(t_0)$ in the neighborhood of the origin in U.

Lemma 6.2 ([98]) *If $0 < p = p_1/p_2 \le 1$, where $p_1 > 0$ and $p_2 > 0$ are positive odd integers, then $|x^p - y^p| \le 2^{1-p}|x - y|^p$.*

Lemma 6.3 ([38]) *For $x_i \in \Re, \ i = 1, ..., n, \ 0 < p \le 1$, then*

$$\left(\sum_{i=1}^n |x_i| \right)^p \le \sum_{i=1}^n |x_i|^p \le n^{1-p} \left(\sum_{i=1}^n |x_i| \right)^p.$$

Lemma 6.4 ([98]) *Let c and d be positive constants and $\gamma(x, y) > 0$ is a real value function. Then*

$$|x|^c |y|^d \le \frac{c\gamma(x, y)|x|^{c+d}}{c + d} + \frac{d\gamma^{-c/d}(x, y)|y|^{c+d}}{c + d}.$$

Lemma 6.5 ([91]) *For a connected undirected graph $\mathcal{G}$, the Laplacian matrix $\mathcal{L}$ has the following property: (1) $\mathcal{L}$ is semi-definite. (2) 0 is a simple eigenvalue of $\mathcal{L}$ and $\mathbf{1}$ is the associated eigenvector. (3) Assuming the eigenvalue of $\mathcal{L}$ is denoted as 0, $\lambda_2, ... \lambda_n$ satisfying $0 \le \lambda_2 \le ... \le \lambda_n$, then the second smallest eigenvalue $\lambda_2 > 0$. Furthermore, if $\mathbf{1}^T x = 0$, then $x^T \mathcal{L} x \ge \lambda_2 x^T x$.*

Consider $y = x^{\frac{p}{q}}$ where p is a positive integer and q is a positive odd integer. If we ignore all the complex roots, then obviously $y = \text{sgn}(x)|x|^{\frac{p}{q}}$ if p is an odd integer; otherwise $y = |x|^{\frac{p}{q}}$.

6.2 PROBLEM FORMULATION

We consider a class of multiple mechanical nonlinear systems

$$M_i\ddot{q}_i + C_i(q_i, \dot{q}_i)\dot{q}_i + D_i(q_i)\dot{q}_i = \tau_i, \tag{6.2}$$

where $q_i = (q_{i1}, \ldots, q_{im})^T \in \Re^m$, $i = 1, \ldots, n$ is the state of the ith system, $\tau_i \in \Re^m$ is the control input vector, $M_i \subset \Re^{m \times m}$ is an inertia matrix, $C_i(q_i, \dot{q}_i)$ is the centripetal and Coriolis matrix and $D_i(q_i)\dot{q}_i$ denotes the friction terms.

Denote $v_i = \dot{q}_i$. The following assumptions are needed for the design of finite-time adaptive consensus controllers.

Assumption 6.2.1 *The graph $\mathcal{G}$ is connected.*

Assumption 6.2.2 *$M_i = \text{diag}\{f_{i1}, \ldots, f_{im}\}$ where f_{ik}, $k = 1, \ldots, m$ are unknown positive constants. $C_i(q_i, v_i)v_i + D_i(q_i)v_i = G_i(q_i, v_i)\theta_i$ where $\theta_i \in R^l$ denotes the vector of unknown parameters and $G_i(q_i, v_i) \in R^{m \times l}$ does not contain unknown parameters. Furthermore, $\|G_i(q_i, v_i)\|_1 \leq \rho P(\|v_i\|)$, where ρ is a known positive constant and $P(\cdot)$ is a polynomial with no constant term, i.e., $P(0) = 0$. f_{ik}, $k = 1, \ldots, m$ and θ_i are in known compact set.*

Remark 6.1

- *For a typical mechanical system, q_i appears in the centripetal force and Coriolis force. For a constant v_0, there exists a constant p such that $\|G_i(q_i, v_0)\| < p$. Taking the two-link robot manipulator as an example, q_i is the joint-variable vector, consisting of joint angles θ_i and a joint offset d_i. The joint angles appear in $G_i(q_i, v_i)$ in terms of $\sin(\theta_i)$ or $\cos(\theta_i)$, and a joint offset d_i is always bounded.*
- *Typical examples of mechanical system satisfying Assumption 6.2.2 include Newton-Euler rigid body and three-link cylindrical robot manipulator, as described in [58].*

6.3 ADAPTIVE FINITE-TIME CONSENSUS CONTROLLER DESIGN

System (6.2) can be written as

$$\dot{q}_i = v_i,$$
$$M_i\dot{v}_i + C_i(q_i, v_i)v_i + D_i(q_i)v_i = \tau_i, \tag{6.3}$$

where $v_i = [v_{i1}, ..., v_{im}]^T$, $i \in \mathcal{V}$ denotes the velocity. The designed control law should ensure that all the agents reach consensus in finite time without additional information. A distributed adaptive control law is designed for each agent such that the positions of all the agents converge to a consensus location based on its neighbor's information. The controller is designed using a "backstepping-like" procedure. In the first step virtual control v_i^* for v_i, $i = 1, ..., n$ is designed such that the position error converges to zero in finite-time. In the second step adaptive controllers and parameter estimators are designed such that v_i converges to v_i^* in finite-time. These finite-time convergence results are summarized in Theorem 6.1. The transient performance of the close-loop adaptive control systems is established and presented in Theorem 6.2, by showing that the position errors and the virtual control errors converge faster than an exponential rate.

Step 1. Let $x_j = [q_{1j} - \delta_{1j}, ..., q_{nj} - \delta_{nj}]^T$, $x = [x_1^T, ..., x_m^T]^T$, $j = 1, ..., m$, where δ_{kj}, $k = 1, ..., n$ are constants denoting the final consensus configuration such that $q_{ij} - q_{kj} = \delta_{ij} - \delta_{kj}$, $i, k \in \mathcal{V}$. Define a Lyapunov function

$$
\begin{aligned}
V_1(t) &= \frac{1}{2} x^T (I_m \otimes \mathcal{L}) x \\
&= \frac{1}{4} \sum_{j=1}^{m} \sum_{i=1}^{n} \sum_{k \in N_i} a_{ik}(q_{ij} - q_{kj} - \delta_{ij} + \delta_{kj})^2.
\end{aligned}
\tag{6.4}
$$

Taking the derivative of V_1 yields

$$
\begin{aligned}
\dot{V}_1 &= \dot{x}^T (I_m \otimes \mathcal{L}) x \\
&= \sum_{j=1}^{m} \sum_{i=1}^{n} \left[\sum_{k \in N_i} a_{ik}(q_{ij} - q_{kj} - \delta_{ij} + \delta_{kj}) \right] \dot{q}_{ij}.
\end{aligned}
$$

Let

$$
e_{ij} = \sum_{k \in N_i} a_{ik}(q_{ij} - q_{kj} - \delta_{ij} + \delta_{kj}).
\tag{6.5}
$$

Then

$$
\dot{V}_1 = \sum_{j=1}^{m} \sum_{i=1}^{n} e_{ij} v_{ij}.
$$

Denoting v_{ij}^* as the virtual control of v_{ij}, we have

$$
\dot{V}_1 = \sum_{j=1}^{m} \sum_{i=1}^{n} e_{ij} v_{ij}^* + \sum_{j=1}^{m} \sum_{i=1}^{n} e_{ij}(v_{ij} - v_{ij}^*).
$$

By choosing the virtual control as

$$
v_{ij}^* = -c_1 e_{ij}^{\frac{2\sigma - 1}{2\sigma + 1}},
\tag{6.6}
$$

where $\sigma \geq 2$ is a positive integer and c_1 is a positive constant to be designed, we get

$$\dot{V}_1 = \sum_{j=1}^{m}\sum_{i=1}^{n} -c_1 e_{ij}^{\frac{4\sigma}{2\sigma+1}} + \sum_{j=1}^{m}\sum_{i=1}^{n} e_{ij}(v_{ij} - v_{ij}^{\star})$$

$$\leq -c_1 \left(\sum_{j=1}^{m}\sum_{i=1}^{n} e_{ij}^2 \right)^{\frac{2\sigma}{2\sigma+1}} + \sum_{j=1}^{m}\sum_{i=1}^{n} e_{ij}(v_{ij} - v_{ij}^{\star}). \tag{6.7}$$

Let $e_j = [e_{1j}, .., e_{nj}]^T$, then $e_j = \mathcal{L}x_j$. Thus $e_j^T e_j = x_j^T \mathcal{L}^T \mathcal{L} x_j = x_j^T \mathcal{L}^2 x_j$. Since $\mathcal{L}$ is a diagonalizable symmetric semi-positive definite matrix, then it is easy to prove that $\mathcal{L}^{1/2}$ is also a symmetric positive semi-definite matrix, and $\mathcal{L}^2 = \mathcal{L}^{1/2}\mathcal{L}\mathcal{L}^{1/2}$. Let $w = L^{1/2}\mathbf{1}$, we get $w^T w = (\mathcal{L}^{1/2}\mathbf{1})^T(\mathcal{L}^{1/2}\mathbf{1}) = \mathbf{1}^T \mathcal{L}\mathbf{1} = 0$. Thus $w = \mathbf{0}$, which yields $w^T x_j = 0$, i.e., $\mathbf{1}^T \mathcal{L}^{1/2} x_j = 0$. Then according to Lemma 6.5

$$e_j^T e_j = (\mathcal{L}^{1/2}x_j)^T \mathcal{L}(\mathcal{L}^{1/2}x_j) \geq \lambda_2 x_j^T \mathcal{L}x_j. \tag{6.8}$$

Let $e = [e_1^T, ..., e_m^T]^T$, then we get

$$\sum_{j=1}^{m}\sum_{i=1}^{n} e_{ij}^2 = e^T e \geq \lambda_2 x^T (I_m \otimes \mathcal{L})x = 2\lambda_2 V_1, \tag{6.9}$$

Step 2. Let $g = \frac{2\sigma-1}{2\sigma+1}$. Invoke the *adding a power integrator* technique as in [97] by defining another Lyapunov function

$$W_i = \frac{1}{2^{1-g}} \sum_{j=1}^{m} f_{ij} \int_{v_{ij}^{\star}}^{v_{ij}} (s^{1/g} - v_{ij}^{\star 1/g})^{2-g} ds, \tag{6.10}$$

and a new variable

$$\xi_{ij} = v_{ij}^{1/g} - v_{ij}^{\star 1/g}, \tag{6.11}$$

where f_{ij} is defined in Assumption 6.2.2. From [98], W_i is positive semi-definite and C^1. Taking the time derivative of W_i, we get

$$\dot{W}_i = \frac{1}{2^{1-g}} \sum_{j=1}^{m} \left[f_{ij}\xi_{ij}^{2-g}\dot{v}_{ij} - (2-g)f_{ij} \int_{v_{ij}^{\star}}^{v_{ij}} (s^{1/g} - v_{ij}^{\star 1/g})^{1-g} ds \sum_{k \in \bar{N}_i} \frac{\partial v_{ij}^{\star 1/g}}{\partial q_{kj}} \dot{q}_{kj} \right],$$

where $\bar{N}_i = N_i \cup \{i\}$. Since

$$\left| \int_{v_{ij}^{\star}}^{v_{ij}} (s^{1/g} - v_{ij}^{\star 1/g})^{1-g} ds \right|$$

$$\leq |v_{ij} - v_{ij}^{\star}||\xi_{ij}|^{1-g}$$

$$= \left| \left(v_{ij}^{1/g} \right)^g - \left(v_{ij}^{\star\,1/g} \right)^g \right| |\xi_{ij}|^{1-g}$$
$$\leq 2^{1-g} |\xi_{ij}^g| |\xi_{ij}^{1-g}| = 2^{1-g} |\xi_{ij}|, \tag{6.12}$$

we have

$$\dot{W}_i \leq \sum_{j=1}^{m} \frac{f_{ij}}{2^{1-g}} \xi_{ij}^{2-g} \dot{v}_{ij} + (2-g) \sum_{j=1}^{m} f_{ij} |\xi_{ij}| \sum_{k \in \bar{N}_i} \left| \frac{\partial v_{ij}^{\star\,1/g}}{\partial q_{kj}} \right| |v_{kj}|$$
$$= \frac{1}{2^{1-g}} \xi_i^T \left[-(C_i(q_i, v_i)v_i + D_i v_i) + \tau_i \right]$$
$$+ (2-g) \sum_{j=1}^{m} f_{ij} |\xi_{ij}| \sum_{k \in \bar{N}_i} \left| \frac{\partial v_{ij}^{\star\,1/g}}{\partial q_{kj}} \right| |v_{kj}|,$$

where

$$\xi_i = [\xi_{i1}^{2-g}, ..., \xi_{im}^{2-g}]^T. \tag{6.13}$$

Let $\hat{\theta}_i$, $i \in \mathcal{V}$, $j = 1, .., m$ denote the estimate of θ_i and define $\tilde{\theta}_i = \hat{\theta}_i - \theta_i$. Then

$$\dot{W}_i \leq \frac{1}{2^{1-g}} \xi_i^T \left(-G_i(q_i, v_i)\hat{\theta}_i + \tau_i \right) + \frac{1}{2^{1-g}} \xi_i^T G_i(q_i, v_i)\tilde{\theta}_i$$
$$+ (2-g)c_1^{1/g} \sum_{j=1}^{m} f_{ij} |\xi_{ij}| \left(\mu|v_{ij}| + \eta \sum_{k \in N_i} |v_{kj}| \right),$$

where $\mu = \max_{\forall i \in \mathcal{V}} \left\{ \sum_{j \in N_i} a_{ij} \right\}$ and $\eta = \max_{\forall i, j \in \mathcal{V}} \{a_{ij}\}$. From Lemma 6.2 and Lemma 6.4, we have

$$\sum_{j=1}^{m} \sum_{i=1}^{n} e_{ij}(v_{ij} - v_{ij}^{\star}) \leq \sum_{j=1}^{m} \sum_{i=1}^{n} 2^{1-g} |e_{ij}| |\xi_{ij}|^g$$
$$\leq \frac{2^{1-g}}{1+g} \sum_{j=1}^{m} \sum_{i=1}^{n} \left(e_{ij}^{1+g} + g\xi_{ij}^{1+g} \right).$$

Define a new Lyapunov function

$$V_2 = V_1 + \sum_{i=1}^{n} W_i. \tag{6.14}$$

Then

$$\dot{V}_2 \leq -c_1 \sum_{j=1}^{m} \sum_{i=1}^{n} e_{ij}^{\frac{4\sigma}{2\sigma+1}} + \frac{2^{1-g}}{1+g} \sum_{j=1}^{m} \sum_{i=1}^{n} \left(e_{ij}^{1+g} + g\xi_{ij}^{1+g} \right)$$
$$+ \frac{1}{2^{1-g}} \sum_{i=1}^{n} \xi_i^T \left(-G_i(q_i, v_i)\hat{\theta}_i + \tau_i \right) + \frac{1}{2^{1-g}} \sum_{i=1}^{n} \xi_i^T G_i(q_i, v_i)\tilde{\theta}_i$$

$$+ (2-g)c_1^{1/g} \sum_{i=1}^{n} \sum_{j=1}^{m} f_{ij}|\xi_{ij}| \left(\mu|v_{ij}| + \eta \sum_{k \in N_i} |v_{kj}| \right). \tag{6.15}$$

Based on Lemmas 6.2–6.4 and (6.11),

$$(2-g)c_1^{1/g} \sum_{i=1}^{n} \sum_{j=1}^{m} f_{ij}|\xi_{ij}| \left(\mu|v_{ij}| + \eta \sum_{k \in N_i} |v_{kj}| \right)$$

$$\leq \sum_{i=1}^{n} \sum_{j=1}^{m} \left(\frac{f_{ij}\mu g + \eta n_i g}{1+g} |e_{ij}|^{1+g} + (2-g)c_1^{1/g} \left(f_{ij}\mu + \right.\right.$$

$$\left.\left. \frac{f_{ij}c_1^2(2-g)c_1^{1/g} + \eta n_i + \eta c_1^2(2-g)c_1^{1/g} + \eta n_i g}{1+g} \right) |\xi_{ij}|^{1+g} \right),$$

where $n_i = \dim N_i$. Since $F_i^{-1} = \mathrm{diag}(1/f_{i1}, ..., 1/f_{im})$, the torque is designed as

$$\tau_i = G(q_i, v_i)\hat{\theta}_i - c_2 \xi_i^*, \tag{6.16}$$

where $\xi_i^* = [\xi_{i1}^{2g-1}, ..., \xi_{im}^{2g-1}]^T$ and c_2 is a positive constant to be chosen. From (6.15) and (6.16), we get

$$\dot{V}_2 \leq -k_1 \sum_{j=1}^{m} \sum_{i=1}^{n} e_{ij}^{\frac{4\sigma}{2\sigma+1}} - k_2 \sum_{j=1}^{m} \sum_{i=1}^{n} \xi_{ij}^{\frac{4\sigma}{2\sigma+1}}$$

$$+ \frac{1}{2^{1-g}} \sum_{i=1}^{n} \xi_i^T G_i(q_i, v_i)\tilde{\theta}_i. \tag{6.17}$$

where

$$k_1 = c_1 - \frac{2^{1-g}}{1+g} - \frac{f_{\max}\mu g + \eta n_i g}{1+g}$$

$$k_2 = c_2 - \frac{g 2^{1-g}}{1+g} - (2-g)c_1^{1/g} \left(f_{\max}\mu + \right.$$

$$\left. \frac{f_{\max}c_1^2(2-g)c_1^{1/g} + \eta n_i + \eta c_1^2(2-g)c_1^{1/g} + \eta n_i g}{1+g} \right), \tag{6.18}$$

with $f_{\max} \geq \max_{j=1,...,m}(f_{1j}, ..., f_{nj})$ being a positive constant known from Assumption 6.2.2. Thus we can find c_1 and c_2 such that $k_1 > 0$ and $k_2 > 0$. Consider the following Lyapunov function

$$V_3 = V_2 + \frac{1}{2} \sum_{i=1}^{n} \tilde{\theta}_i^T \Gamma_i^{-1} \tilde{\theta}_i, \tag{6.19}$$

where Γ_i is a positive diagonal matrix. Then

$$\dot{V}_3 \leq -k_1 \sum_{j=1}^{m} \sum_{i=1}^{n} e_{ij}^{\frac{4\sigma}{2\sigma+1}} - k_2 \sum_{j=1}^{m} \sum_{i=1}^{n} \xi_{ij}^{\frac{4\sigma}{2\sigma+1}}$$

$$+ \sum_{i=1}^{n} \tilde{\theta}_i^T \Gamma_i^{-1} \left(\dot{\hat{\theta}}_i + \frac{1}{2^{1-g}} \Gamma_i G_i(q_i, v_i)^T \xi_i \right).$$

The parameter update law for θ_i is

$$\dot{\hat{\theta}}_{ik} = \text{Proj}\left(\beta_i(k), \hat{\theta}_{ik} \right), \quad k = 1, ..., l, \tag{6.20}$$

where $\beta_i(k)$ is the kth element of $\beta_i = -\frac{1}{2^{1-g}} \Gamma_i G_i(q_i, v_i)^T \xi_i$. The operator $\text{Proj}(\cdot, \cdot)$ is a Lipschitz continuous projection algorithm in [95] which is defined as follows:

$$\text{Proj}(a, \hat{b}) = \begin{cases} a & \text{if} \quad \mu(\hat{b}) \leq 0; \\ a & \text{if} \quad \mu(\hat{b}) \geq 0 \text{ and } \mu'(\hat{b})a \leq 0; \\ (1 - \mu(\hat{b}))a & \text{if} \quad \mu(\hat{b}) > 0 \text{ and } \mu'(\hat{b})a > 0, \end{cases} \tag{6.21}$$

where $\mu(\hat{b}) = \frac{\hat{b}^2 - b_M^2}{\epsilon^2 + 2\epsilon b_M}$, $\mu'(\hat{b}) = \frac{\partial \mu(\hat{b})}{\hat{b}}$, ϵ is an arbitrarily small positive constant, b_M is a positive constant satisfying $|b| < b_M$.

Then from (6.17), (6.19), (6.20) and based on the property of the parameter projection in [56], we get

$$\dot{V}_3 \leq -k_1 \sum_{j=1}^{m} \sum_{i=1}^{n} e_{ij}^{\frac{4\sigma}{2\sigma+1}} - k_2 \sum_{j=1}^{m} \sum_{i=1}^{n} \xi_{ij}^{\frac{4\sigma}{2\sigma+1}}. \tag{6.22}$$

6.4   STABILITY ANALYSIS

We are now at the position to state our first result in the following theorem.

Theorem 6.1 *Consider the undirected leaderless multi-agent uncertain systems (6.2), under the control of the distributed adaptive controllers (6.16) with parameter estimators in (6.20). If Assumptions 6.2.1 and 6.2.2 are satisfied, then the positions of the group of mechanical systems will reach consensus with specified configurations δ_{ij} in finite time T satisfying*

$$T \leq \frac{\varrho_1^{1-\kappa}}{k_x(1-\kappa)} + \frac{V_3(t_0) - \varrho_1}{d_{v_3}}, \tag{6.23}$$

where κ, k_x, ϱ_1, and d_{v_3} are computable.

Proof. From (6.22) we get $\dot{V}_3 \leq 0$. Therefore e_{ij} and $\tilde{\theta}_i$ are bounded and it can be easily checked that $\lim_{t \to \infty} e_{ij} = 0$. Since we cannot use $\dot{V}_3 + cV_3^\alpha \leq 0$ to establish the finite-time stability of e_{ij} and ξ_{ij} due to the existence of the terms $\tilde{\theta}_i^T \tilde{\theta}_i$, we will prove that there exist two positive constants k_x and $0 < \kappa < 1$ such that $\dot{V}_2 + k_x V_2^\kappa \leq 0$ which enable us to show that the systems are finite-time stable.

By using the parameter projection operation and based on Assumption 6.2.2, we know there exists a positive constant S_i such that $\|\tilde{\theta}_i\| \leq S_i$. Then from (6.10) and (6.12) we have

$$W_i \leq \sum_{j=1}^{m} f_{ij} \xi_{ij}^2 \leq f_{\max} \sum_{j=1}^{m} \xi_{ij}^2. \tag{6.24}$$

Thus from (6.9) we obtain

$$V_2 \leq \frac{1}{2\lambda_2} \sum_{i=1}^{n} \sum_{j=1}^{m} e_{ij}^2 + f_{\max} \sum_{i=1}^{n} \sum_{j=1}^{m} \xi_{ij}^2. \tag{6.25}$$

Suppose $P(\cdot)$ in Assumption 6.2.2 is

$$P(x) = p_1 x + p_2 x^2 + \ldots + p_h x^h, \tag{6.26}$$

where $h \geq 2$ is an integer and p_k, $k = 1, \ldots, h$ are positive constants. From (6.11)

$$|v_{ij}| \leq (|\xi_{ij}| + |v_{ij}^\star|^{1/g})^g \leq |\xi_{ij}|^g + c_1 |e_{ij}|^g. \tag{6.27}$$

Define a compact set

$$\Xi = \{\xi_{ij}, e_{ij} : |e_{ij}| < r_{ij}, |\xi_{ij}| < d_{ij}\}, \tag{6.28}$$

in the neighborhood of $e_{ij} = 0$, $\xi_{ij} = 0$ where r_{ij} and d_{ij} are positive constants chosen to be

$$r_{ij} = \left(\frac{1}{2c_1} \sqrt{\frac{1}{m}} \right)^{1/g}, \quad d_{ij} = \left(\frac{1}{2} \sqrt{\frac{1}{m}} \right)^{1/g}. \tag{6.29}$$

From (6.27) it is easy to check that in this set we have $\|v_i\| \leq 1$. Thus from Assumption 6.2.2 we get $\|G_i(q_i, v_i)\|_1 \leq \rho h p_{\max} \|v_i\|$ where $p_{\max} = \max(p_1, \ldots, p_h)$. Furthermore, $|G_i(q_i, v_i)(j, k)| < \rho h p_{\max} |v_{ij}|$, $k = 1, \ldots, l$ since $|v_{ij}| \leq 1$. Then from (6.13), (6.17), and (6.27),

$$\frac{1}{2^{1-p}} \sum_{i=1}^{n} \xi_i^T G_i(q_i, v_i) \tilde{\theta}_i \leq \frac{1}{2^{1-p}} \sum_{i=1}^{n} \left\| |\xi_i|^T |G_i(q_i, v_i)| \right\| \|\tilde{\theta}_i\|$$

$$\leq \frac{S_i}{2^{1-p}} \sum_{i=1}^{n} \sum_{j=1}^{m} \rho h p_{\max} |\xi_{ij}^{2-g}| \left(|\xi_{ij}|^g + c_1 |e_{ij}|^g \right)$$

$$= k_3 \sum_{i=1}^{n} \sum_{j=1}^{m} e_{ij}^2 + k_4 \sum_{i=1}^{n} \sum_{j=1}^{m} \xi_{ij}^2, \tag{6.30}$$

where $|\xi_i|$ and $|G_i(q_i, v_i)|$ are taken element-wisely, and k_3 and k_4 are

$$k_3 = \frac{S_{\max} \rho h p_{\max} g}{2^{2-p}}, \quad k_4 = \frac{S_{\max} \rho h p_{\max}}{2^{1-p}} \left(1 + \frac{2-g}{2c_1^2} \right). \tag{6.31}$$

in which $S_{\max}$ is defined as $S_{\max} = \max(S_1, ..., S_n)$. Thus from (6.17) we get

$$\dot{V}_2 \leq -\frac{k_1}{2} \sum_{j=1}^{m} \sum_{i=1}^{n} e_{ij}^{\frac{4\sigma}{2\sigma+1}} - \frac{k_2}{2} \sum_{j=1}^{m} \sum_{i=1}^{n} \xi_{ij}^{\frac{4\sigma}{2\sigma+1}} \tag{6.32}$$

$$+ \sum_{j=1}^{m} \sum_{i=1}^{n} \left(k_3 e_{ij}^2 - \frac{k_1}{2} e_{ij}^{\frac{4\sigma}{2\sigma+1}} \right) + \sum_{j=1}^{m} \sum_{i=1}^{n} \left(k_4 \xi_{ij}^2 - \frac{k_2}{2} \xi_{ij}^{\frac{4\sigma}{2\sigma+1}} \right).$$

Now we will establish the finite-time stability in two cases. In the first case we will prove that e_{ij} and ξ_{ij} will converge to zero in finite-time if $e_{ij}(t_0) \in \Xi$ and $\xi_{ij}(t_0) \in \Xi$. In the second case we will show that e_{ij} and ξ_{ij} will converge to Ξ within finite-time and thus converge to zero in finite-time if $e_{ij}(t_0) \notin \Xi$ and/or $\xi_{ij}(t_0) \notin \Xi$.

Case 1: The initial conditions of e_{ij} and ξ_{ij} satisfy

$$e_{ij}(t_0) \in \Xi, \ \xi_{ij}(t_0) \in \Xi.$$

From (6.18) and (6.31) we can choose the design parameters c_1 and c_2 such that $k_1 > 2k_3$ and $k_2 > 2k_4$. Since $|e_{ij}| < 1$ and $|\xi_{ij}| < 1$, it is not hard to see that $\sum_{j=1}^{m} \sum_{i=1}^{n} \left(k_3 e_{ij}^2 - \frac{k_1}{2} e_{ij}^{\frac{4\sigma}{2\sigma+1}} \right) + \sum_{j=1}^{m} \sum_{i=1}^{n} \left(k_4 \xi_{ij}^2 - \frac{k_2}{2} \xi_{ij}^{\frac{4\sigma}{2\sigma+1}} \right) < 0$ if we choose c_1 and c_2 in this way. Thus we have

$$\dot{V}_2 \leq -\frac{k_1}{2} \sum_{j=1}^{m} \sum_{i=1}^{n} e_{ij}^{\frac{4\sigma}{2\sigma+1}} - \frac{k_2}{2} \sum_{j=1}^{m} \sum_{i=1}^{n} \xi_{ij}^{\frac{4\sigma}{2\sigma+1}}. \tag{6.33}$$

From (6.25) and Lemma 6.3 we get $V_2^{\kappa} \leq \frac{1}{2^{\kappa} \lambda_2^{\kappa}} \sum_{i=1}^{n} \sum_{j=1}^{m} e_{ij}^{\frac{4\sigma}{2\sigma+1}} + f_{\max}^{\kappa} \sum_{i=1}^{n} \sum_{j=1}^{m} \xi_{ij}^{\frac{4\sigma}{2\sigma+1}}$

where $\kappa = \frac{2\sigma}{2\sigma+1} \in (0,1)$. Hence

$$\dot{V}_2 + k_x V_2^{\kappa} \leq 0, \tag{6.34}$$

where $k_x < \min\left\{ 2^{\kappa-1} \lambda_2^{\kappa} k_1, \frac{k_2}{2 f_{\max}^{\kappa}} \right\}$ is a positive constant. Since $0 < g < 1$, then based on Lemma 6.1, V_2 will converge to zero in a finite time T_1 which satisfies

$$T_1 \leq \frac{V_2(t_0)^{1-\kappa}}{k_x(1-\kappa)}. \tag{6.35}$$

Case 2: The initial values of e_{ij} and/or ξ_{ij} are outside the set Ξ. For the period of time when e_{ij} and/or ξ_{ij} are outside of the set Ξ, from (6.22), we know

$$\dot{V}_3 \leq -d_{v3}, \tag{6.36}$$

where $d_{v3} \geq \min\left(k_1 \sum_{i=1}^{n} \sum_{j=1}^{m} r_{ij}^{\frac{4\sigma}{2\sigma+1}}, k_2 \sum_{i=1}^{n} \sum_{j=1}^{m} d_{ij}^{\frac{4\sigma}{2\sigma+1}} \right)$ is a positive constant.

To determine the time T_2 that e_{ij} and ξ_{ij} reach the set Ξ from the initials, we have to calculate a ϱ_1 such that $V_3 \geq \varrho_1$ when $e_{ij}(t) \notin \Xi$ and/or $\xi_{ij}(t) \notin \Xi$. Then T_2 will not be longer than $(V_3(t_0) - \varrho_1)/d_{v_3}$. From (6.14) and (6.19), such a ϱ_1 is calculated as

$$\varrho_1 = \min \left\{ V_2|_{(e_{ij}=r_{ij}, \xi_{ij}=0)}, \ \left(\sum_{i=1}^{n} W_i \right) \bigg|_{(\xi_{ij}=d_{ij}, |e_{ij}|<r_{ij})} \right\}.$$

From (6.4) we know

$$V_2|_{(e_{ij}=r_{ij}, \xi_{ij}=0)} = V_1|_{e_{ij}=r_{ij}}$$

$$\geq \frac{\zeta}{4\eta^2} \sum_{j=1}^{m} \sum_{i=1}^{n} \sum_{k \in N_i} \eta^2 (q_{ij} - q_{kj} - \delta_{ij} + \delta_{kj})^2$$

$$\geq \frac{\zeta}{4\eta^2 n_{\max}} \sum_{j=1}^{m} \sum_{i=1}^{n} \left[\sum_{k \in N_i} a_{ik}(q_{ij} - q_{kj} - \delta_{ij} + \delta_{kj}) \right]^2$$

$$= \frac{\zeta}{4\eta^2 n_{\max}} \sum_{j=1}^{m} \sum_{i=1}^{n} r_{ij}^2, \tag{6.37}$$

where $\zeta = \min_{\forall i,j \in \mathcal{V}}\{a_{ij}\}$, $n_{\max} = \max\{n_1, ..., n_n\}$. $\eta = \max_{\forall i,j \in \mathcal{V}}\{a_{ij}\}$, $n_i = \dim\{N_i\}$. It can be checked that

$$\int_{v_{ij}^\star}^{v_{ij}} (s^{1/g} - v_{ij}^{\star 1/g})^{2-g} ds \geq \frac{2\sigma - 1}{2^{\frac{(1-g)(2-g)}{g}}(4\sigma+2)} \left| v_{ij} - v_{ij}^\star \right|^{\frac{4\sigma+2}{2\sigma-1}}$$

$$\geq \frac{2\sigma - 1}{2^{\frac{(1-g)(2-g)}{g}}(4\sigma+2)} \left((d_{ij} + r_{ij})^g - r_{ij}^g \right)^{\frac{4\sigma+2}{2\sigma-1}} := s_{ij}. \tag{6.38}$$

This is because for a fixed ξ_{ij}, $v_{ij} - v_{ij}^\star$ decreases monotonously as e_{ij} increases. Thus from (6.10)

$$\left(\sum_{i=1}^{n} W_i \right) \bigg|_{\xi_{ij}=d_{ij}, |e_{ij}|<r_{ij}} = \sum_{i=1}^{n} \sum_{j=1}^{m} \frac{f_{ij}}{2^{1-g}} s_{ij} \geq \frac{f_{\min}}{2^{1-g}} \sum_{i=1}^{n} \sum_{j=1}^{m} s_{ij},$$

where $f_{\min} \leq \min_{\forall i,j \in \mathcal{V}} f_{ij}$ is a positive constant known from Assumption 6.2.2. So with ϱ_1, e_{ij} and ξ_{ij} will reach the set within a time T_2 which satisfies

$$T_2 \leq \frac{V_3(t_0) - \varrho_1}{d_{v_3}}. \tag{6.39}$$

After reaching the set, e_{ij} and ξ_{ij} will stay in this set forever because $\dot{V}_2 < 0$ in this set, and it will take a time less than $T_3 = \dfrac{\varrho_1^{1-\kappa}}{k_x(1-\kappa)}$ to arrive at the origin. Thus in this case the system will be finite time stable with T satisfying (6.23). In summary, the closed-loop system is globally finite-time convergent. This completes the proof.

Remark 6.2 *From (6.25) and (6.36) we know k_x and d_{v_3} will be larger if we choose larger controller parameters c_1 and c_2. Since κ is a constant, thus from (6.18) and (6.22) we know (6.23) can be adjusted arbitrarily small, but it will result in larger control effort.*

Remark 6.3 *From the proof of Theorem 6.1 we can see that local finite-time stability and global asymptotical stability yield global finite-time stability.*

6.5   TRANSIENT PERFORMANCE ANALYSIS

As mentioned in the introduction, for adaptive control of nonlinear systems it is very difficult to establish exponential convergence rate. Nevertheless we explore the transient performance of the resulting closed-loop adaptive control system here. The following theorem is established.

Theorem 6.2 *Tracking errors e_{ij} and virtual control errors ξ_{ij} converge to the origin with a speed faster than an exponential rate when $e_{ij} \in \Xi$ and $\xi_{ij} \in \Xi$.*

Proof. Based on (6.34) which holds when $(e_{ij}, \xi_{ij}) \in \Xi$, we get $\dot{V}_2 \leq -k_x V_2^{\kappa} \leq 0$. Thus $V_2 \leq V_2(t_0)$, and we can find a positive constant k_p such that

$$\dot{V}_2 + k_p V_2 \leq 0, \tag{6.40}$$

where $k_p = k_x^{1/\kappa} V_2(t_0)^{\frac{\kappa-1}{\kappa}}$. Therefore

$$V_2(t) \leq V_2(t_0)e^{-k_p t}. \tag{6.41}$$

From (6.37) we get $V_2 \geq \dfrac{\zeta}{4\eta^2 n_{\max}} \displaystyle\sum_{j=1}^{m}\sum_{i=1}^{n} e_{ij}^2$. Thus the tracking errors satisfy

$$\|e(t)\|^2 = \sum_{j=1}^{m}\sum_{i=1}^{n} e_{ij}(t)^2 \leq \frac{4\eta^2 n_{\max}}{\zeta} V_2(t_0)e^{-k_p t}. \tag{6.42}$$

From (6.38) we have

$$\sum_{j=1}^{m}\sum_{i=1}^{n} \left| v_{ij} - v_{ij}^{\star} \right|^{\frac{4\sigma+2}{2\sigma-1}} \leq \vartheta_0 V_2(t_0)e^{-k_p t}, \tag{6.43}$$

where $\vartheta_0 = \dfrac{2^{\frac{(1-g)(2-g)}{g}}(4\sigma+2)}{2\sigma-1}$. Based on (6.6), (6.42), and (6.43), we get

$$|v_{ij}| \leq |v_{ij} - v_{ij}^{\star}| + |v_{ij}^{\star}| \leq \vartheta_1 e^{-k_{v1} t} + \vartheta_2 e^{-k_{v2} t}, \tag{6.44}$$

where $\vartheta_1 = (\vartheta_0 V_2(t_0))^{\frac{2\sigma-1}{4\sigma+2}}$, $k_{v1} = k_p \dfrac{2\sigma-1}{4\sigma+2}$, $k_{v2} = \dfrac{k_p g}{2}$, $\vartheta_2 = c_1\left(\dfrac{4\eta^2 n_{\max}}{\zeta} V_2(t_0)\right)^{\frac{g}{2}}$. From (6.44) we get $|v_{ij}| \leq \vartheta_1 + \vartheta_2$, and $|v_{ij}^{\star}| < \vartheta_2$.

According to the differential mean value theorem, there exists a ψ with $|\psi| <$ $\vartheta_1 + \vartheta_2$, such that $|v_{ij}^{1/g} - v_{ij}^{\star 1/g}| = \frac{1}{g}|\psi|^{\frac{1-g}{g}}|v_{ij} - v_{ij}^{\star}|$. Since function $y = |x|^{\frac{1-g}{g}}$ is monotonously decreasing or increasing when $x < 0$ or $x \geq 0$ respectively, thus there exists a positive constant k_σ, such that

$$|v_{ij}^{1/g} - v_{ij}^{\star 1/g}| < k_\sigma |v_{ij} - v_{ij}^{\star}|, \tag{6.45}$$

where $k_\sigma = \frac{1}{g}(\vartheta_1 + \vartheta_2)^{\frac{1-g}{g}}$. Then from Lemma 6.3, (6.43), and (6.45) we obtain

$$\|\xi(t)\|^2 = \sum_{j=1}^{m}\sum_{i=1}^{n}\xi_{ij}(t)^2 \leq \sum_{j=1}^{m}\sum_{i=1}^{n}k_\sigma^2|v_{ij} - v_{ij}^{\star}|^2$$

$$\leq k_\sigma^2\sqrt{mn}\Big(\sum_{j=1}^{m}\sum_{i=1}^{n}|v_{ij} - v_{ij}^{\star}|^4\Big)^{\frac{1}{2}}$$

$$\leq k_\sigma^2\sqrt{mn}\Big(\sum_{j=1}^{m}\sum_{i=1}^{n}|v_{ij} - v_{ij}^{\star}|^{\frac{4\sigma+2}{2\sigma-1}}\Big)^{\frac{1}{2}\frac{8\sigma-4}{4\sigma+2}} \leq k_\xi e^{-k_s t}, \tag{6.46}$$

where $k_\xi = k_\sigma^2\sqrt{mn}\Big(\vartheta_0 V_2(t_0)\Big)^{\frac{4\sigma-2}{4\sigma+2}}$ and $k_s = k_p\frac{4\sigma-2}{4\sigma+2}$. This completes the proof.

Remark 6.4

- *From (6.42) and (6.46) we know the tracking errors and the virtual control errors converge with exponential rate at least and will converge to the origin within finite time if $(e_{ij}, \xi_{ij}) \in \Xi$.*
- *On the other hand, if $e_{ij}(t_0) \notin \Xi$ and/or $\xi_{ij}(t_0) \notin \Xi$, both e_{ij} and ξ_{ij} will enter Ξ within T_2 defined in (6.39). From (6.9) we know $\|e(t_0)\| \geq 2\lambda_2 V_1(t_0)$. However at $t = T_2$, $\|e\| < mnr_{ij}^2$. Thus for $t \in [t_0, T_2]$, $\|e(t)\|$ drops from $2\lambda_2 V_1(t_0)$ to mnr_{ij}^2. For ξ_{ij}, suppose $v_{ij}(t_0) = 0$, then for $t = t_0$, $\|\xi\| \geq 2c_1^{2/g}V_1(t_0)$, and for $t = T_2$, $\|\xi(T_2)\| \leq mnd_{ij}^2$. As discussed in Remark 6.2, T_2 can be made arbitrarily small by adjusting the control parameters. Therefore, any pre-specified convergence time can be met by choosing appropriate design parameters for any initial values of tracking errors and virtual control errors.*
- *If the upper bound of θ is unknown, then from (6.22) we can still choose the control parameters c_1 and c_2 by incorporating $|\tilde{\theta}| \leq \sqrt{2V_3(t_0)}$ such that $\frac{1}{2^{1-g}}\sum_{i=1}^{n}\xi_i^T G_i(q_i, v_i)\tilde{\theta}_i \leq \frac{k_1}{2}\sum_{j=1}^{m}\sum_{i=1}^{n}e_{ij}^{\frac{4\sigma}{2\sigma+1}} + \frac{k_2}{2}\sum_{j=1}^{m}\sum_{i=1}^{n}\xi_{ij}^{\frac{4\sigma}{2\sigma+1}}$. This will ensure that e_{ij} and ξ_{ij} converge to Ξ and then converge to the origin within finite time. However this is not a global result since the initial conditions of system states must be incorporated into the controllers.*

In order to consider the rendezvous problem, we now define the following set of error variables

$$z_{ij} = \sum_{k \in N_i} a_{ik}\left(q_{ij} - q_{kj} - \delta_{ij} + \delta_{kj}\right) + b_i(q_{ij} - \delta_{ij} - Y_j), \qquad (6.47)$$

where $Y = \{Y_1, ..., Y_m\}$ is a rendezvous location which is only available to part of the agents. Let $z_j = [z_{1j}, ..., z_{nj}]^T$, Then we have

$$z_j = (\mathcal{L} + \mathcal{B})(\underline{x}_j - \underline{Y}_j), \qquad (6.48)$$

where $z_j = [z_{1j}, ..., z_{nj}]^T$, $\underline{x}_j = [q_{1j} - \delta_{1j}, ..., q_{nj} - \delta_{nj}]^T$ and $\underline{Y}_j = [Y_j, ..., Y_j]^T \in R^n$. Define a Lyapunov function

$$V_4 = \frac{1}{2} z^T \left(I_m \otimes (\mathcal{L} + \mathcal{B})\right)^{-1} z, \qquad (6.49)$$

where $z = [z_1^T, ..., z_m^T]^T$. Taking the derivative of V_4 yields

$$\dot{V}_4 = z^T v = \sum_{i=1}^{n} \sum_{j=1}^{m} z_{ij} v_{ij},$$

where $v = [v_1^T, ..., v_m^T]^T$ with $v_j = [v_{1j}, ..., v_{nj}]$, $j = 1, ..., m$. Choose the virtual control of v_{ij} as

$$v_{ij}^* = -c_3 z_{ij}^g. \qquad (6.50)$$

Then we get

$$\dot{V}_4 = -c_3 \sum_{i=1}^{n} \sum_{j=1}^{m} z_{ij}^{1+g} + \sum_{i=1}^{n} \sum_{j=1}^{m} (v_{ij} - v_{ij}^*).$$

Also from (6.49) we know

$$z^T z = \sum_{i=1}^{n} \sum_{j=1}^{m} z_{ij}^2 \geq \frac{2}{\lambda_{\max}\left(I_m \otimes (\mathcal{L} + \mathcal{B})\right)^{-1}} V_4. \qquad (6.51)$$

Define a new Lyapunov function

$$V_5 = V_4 + \sum_{i=1}^{n} \Delta_i, \qquad (6.52)$$

where

$$\Delta_i = \frac{1}{2^{1-g}} \sum_{j=1}^{m} f_{ij} \int_{v_{ij}^*}^{v_{ij}} (s^{1/g} - v_{ij}^{*\,1/g})^{2-g} ds \qquad (6.53)$$

and design the controller τ_i as

$$\tau_i = -G_i(q_i, v_i)\hat{\theta}_i - c_2\chi_i^*, \tag{6.54}$$

where $\chi_{ij} = v_{ij}^{1/g} - v_{ij}^{*\,1/g}$ and $\chi_i^* = [\chi_{i1}^{2g-1}, ..., \chi_{im}^{2g-1}]$. Then

$$
\dot{V}_5 \leq -k_1 \sum_{j=1}^{m}\sum_{i=1}^{n} z_{ij}^{\frac{4\sigma}{2\sigma+1}} - k_2 \sum_{j=1}^{m}\sum_{i=1}^{n} \chi_{ij}^{\frac{4\sigma}{2\sigma+1}}
$$

$$
+ \frac{1}{2^{1-g}} \sum_{i=1}^{n} \chi_i^T G_i(q_i, v_i)\tilde{\theta}_i \tag{6.55}
$$

Thus by defining

$$V_6 = V_5 + \frac{1}{2}\sum_{i=1}^{n}\tilde{\theta}_i^T \Gamma^{-1}\tilde{\theta}_i, \tag{6.56}$$

and choosing the parameter update law

$$\dot{\hat{\theta}}_{ik} = \mathrm{Proj}\left(\gamma_i(k), \hat{\theta}_{ik}\right),\ k = 1, ..., l, \tag{6.57}$$

where $\gamma_i = -\dfrac{1}{2^{1-g}}\Gamma_i G_i(q_i, v_i)^T \chi_i$ and $\gamma_i(k)$ is the kth element of γ_i, we get

$$
\dot{V}_6 \leq -k_1 \sum_{j=1}^{m}\sum_{i=1}^{n} z_{ij}^{\frac{4\sigma}{2\sigma+1}} - k_2 \sum_{j=1}^{m}\sum_{i=1}^{n} \chi_{ij}^{\frac{4\sigma}{2\sigma+1}}. \tag{6.58}
$$

Then from (6.58) the following corollary is established.

Corollary 6.1 *Consider multi-agent systems (6.2) under the control of distributed adaptive controllers (6.50) and (6.54) with parameter estimators (6.57). If Assumptions 6.2.1 and 6.2.2 are satisfied, then all the positions of the agents converge to Y with specified configurations δ_{ij} within finite time T satisfying*

$$T \leq \frac{\varrho_2^{1-\kappa}}{k_y(1-\kappa)} + \frac{V_6(t_0) - \varrho_2}{d_{v_6}}. \tag{6.59}$$

Proof. Based on (6.58) and following the similar procedure from (6.24) to (6.39), the conclusion can be established.

6.6   SIMULATION RESULTS

In this section we use four cylindrical robot arms chosen from Section 3.2 in [58] to demonstrate the effectiveness of our proposed finite-time control schemes. The robot parameters are: $M = [m_1\ 0; 0\ m_2]$, $C = [\cos(q_1)\ \mathfrak{c}_1\dot{q}_2; \mathfrak{c}_2\dot{q}_2\ \sin(q_2)]$, $D =$

$[d_1 \ 0; 0 \ d_2]$ with $q = [q_1 \ q_2]^T$, $m_1 = m_2 = 5$, $\mathfrak{c}_1 = \mathfrak{c}_2 = 2$, $d_1 = d_2 = 3$. The control parameters are chosen as $c_1 = 5$, $c_2 = 13$, $c_{11} = c_{21} = c_{31} = 3$, $c_{12} = c_{22} = c_{32} = 10$, $\Gamma_i = \Gamma = I_4$. The initial values of the estimates are selected as 60% of their true values, respectively. The connection graph is shown in Figure 6.1. The initial positions of the robots are at $(20, 54)$, $(4, 20)$, $(44, 1)$, and $(-2, -29)$ respectively. Figure 6.2 shows the positions of the robots. For rendezvous seeking, Figure 6.1 is also the connection graph and only robot 1 has access to the rendezvous point, which is given to be $(43, 65)$. The initial positions are at $(0, 4)$, $(4, 20)$, $(14, 1)$, and $(12, -29)$. The results on positions of the robots are shown in Figure 6.3.

To see the effects of control parameters on the convergence time, we consider two group of parameters: $c_1 = 20$, $c_2 = 5$ and $c_1 = 5$, $c_2 = 5$, respectively. In Figures 6.4 and 6.5, $\|e(t)\|$ and the torque of Robot 1 are presented. These results confirm our discussions in Remark 6.2.

We also make a comparison on convergence time between our proposed scheme and a typical non-finite-time-based adaptive scheme that was proposed in [77], where the controller is given as

$$v_i^\star = -c_1 e_i, \tau_i = -c_2(v_i - v_i^\star) + Y_i(q_i, v_i)\hat{\Theta}_i,$$
$$\dot{\hat{\Theta}}_i = -\Lambda_i Y_i(q_i, v_i)^T e_i, \tag{6.60}$$

where $v_i^\star$ is the virtual control for the velocity v_i, $\hat{\Theta}_i$ is the estimate of unknown system parameter Θ_i, and e_i is the position error defined similarly to (6.5). Both controllers are applied to the same group of robots with the same unknown parameters mentioned above for leaderless consensus. Figure 6.6 shows the comparison results with the same controller parameters $c_1 = 5$ and $c_2 = 5$, the same initial positions and the same initial parameter estimates. It can be observed that $\|e(t)\|$ converge to 0.0001 in 0.06s with our proposed controllers whereas it takes 0.25s to converge to 0.0024 when the controller in [77] is used.

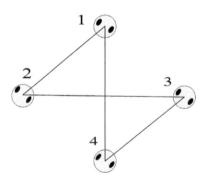

FIGURE 6.1   Undirected graph for leaderless consensus and rendezvous seeking.

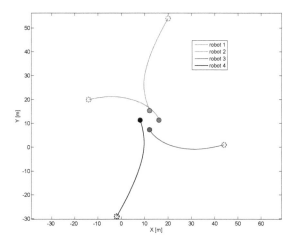

FIGURE 6.2 Positions of four mechanical robots in leaderless consensus.

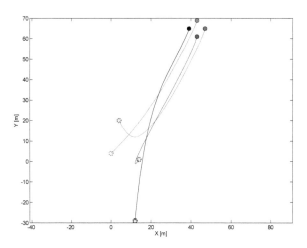

FIGURE 6.3 Positions of four mechanical robots in location seeking.

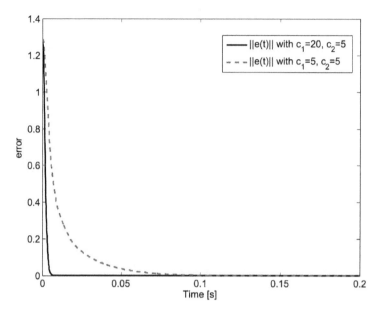

FIGURE 6.4 Convergence rates with different control parameters for the case of leaderless consensus. Black solid line: $c_1 = 20$, $c_2 = 5$. Green dash line: $c_1 = 5$, $c_2 = 5$.

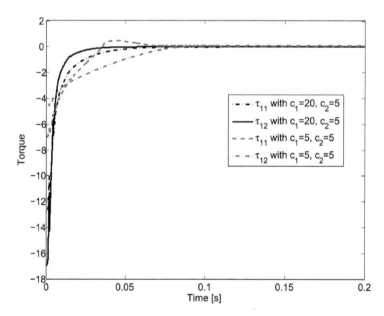

FIGURE 6.5 Robot 1's torque with different control parameters for the leaderless consensus. Black lines: $c_1 = 20$, $c_2 = 5$. Green lines: $c_1 = 5$, $c_2 = 5$.

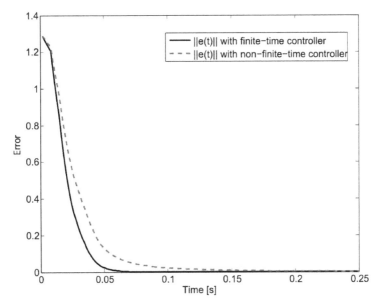

FIGURE 6.6 Convergence time and rates with different controllers for leaderless consensus. Green dash line: non-finite-time controllers in [77]; Black solid line: finite-time controllers

6.7 NOTES

In this chapter, we have investigated the finite-time consensus control for a group of nonlinear mechanical systems with parametric uncertainties. New continuous distributed controllers are proposed for the multi-agent systems. For the leaderless multi-agent systems, it is shown that the states of the mechanical systems can reach a consensus within finite time. We also prove that with our control schemes all the systems can reach a static rendezvous location in finite time under an directed graph when only part of the agents have accesses to the rendezvous location. Transient performances in terms of convergence rates and time are also analyzed and established.

Section III

Consensus Control with Event-Triggered Communication

7 Distributed Event-Triggered Adaptive Control: Leader-Following Case

In previous chapters, a series of distributed adaptive consensus control schemes have been presented. However, the implementation of these schemes requires continuous communication among the agents. As discussed in Chapter 1, this type of communication will consume considerable bandwidth resources, which poses a huge challenge to the agents with limited resources. Event-triggered control is a type of control system strategy in which the control actions are initiated or triggered by specific events or conditions rather than occurring at fixed, periodic intervals. In event-triggered control systems, the controller only takes action when certain criteria or thresholds are met. These criteria could be based on changes in the system's state, the occurrence of specific events, or the satisfaction of certain conditions. The main idea behind event-triggered control is to reduce the frequency of control updates and, consequently, save bandwidth or computational resources in systems where continuous control is not necessary. This approach is often used in real-time systems, cyber-physical systems, and networked control systems to optimize the use of resources. Event-based consensus becomes a significant and hot research topic in recent years. A plenty of representative works in this area have been reported, see [19, 118, 147, 170] for instance. Note that the aforementioned results are mainly established for linear multi-agent systems. However, physical systems are usually nonlinear with system uncertainties in practice.

This chapter investigates the distributed adaptive consensus tracking problem for a class of uncertain high-order nonlinear multi-agent systems with directed graph and event-triggered communication. For each agent, a group of triggering conditions to broadcast its state information is designed, which is only dependent on its local state changing rate. Hence each agent needs no longer monitor its neighbors' states continuously as required in some existing results. Then a totally distributed consensus tracking control scheme based on backstepping technique is proposed. The main challenge is that the virtual controllers designed in each agent will contain piecewise continuous state signal received from its neighbors, since event-based communication mechanism is adopted. The non-differentiable virtual control signals make the recursive design steps of backstepping difficult to proceed. To overcome this obstacle, distributed adaptive backs tepping-based consensus controllers are firstly designed with continuous communication among the

DOI: 10.1201/9781003394372-7

agents. For the event-based communication case, a constructive method is adopted to design the distributed adaptive controllers, where the partial derivatives in previously designed virtual controllers and the neighbors' states collected at the triggering instants are utilized. It is shown that uniform boundedness of all the closed-loop signals can be ensured, while Zeno behavior is ruled out. Besides, it is worth emphasizing that the consensus tracking performance in the mean square sense can be improved by properly adjusting the design parameters.

7.1 PROBLEM FORMULATION

In this chapter, we consider a group of nth-order nonlinear multi-agent systems modeled as follows.

$$
\begin{aligned}
\dot{x}_{i,q} &= x_{i,q+1}, \quad q = 1, 2, \ldots, n-1, \\
\dot{x}_{i,n} &= u_i + \psi_i(x_i) + \varphi_i(x_i)^T \theta_i, \\
y_i &= x_{i,1}, \quad i = 1, 2, \ldots, N,
\end{aligned}
\tag{7.1}
$$

where $x_i = [x_{i,1}, x_{i,2}, \ldots, x_{i,n}]^T \in \Re^n$, $y_i \in \Re$ and $u_i \in \Re$ are the state vector, output and input of the ith agent, respectively. $\theta_i \in \Re^{p_i}$ is a vector of unknown constants. $\psi_i(x_i) \in \Re$ and $\varphi_i(x_i) : \Re^n \to \Re^{p_i}$ are column vectors of known continuous nonlinear functions.

The desired trajectory for all agent outputs is characterized by a bounded time varying function $y_r(t)$. We now use $\mu_i = 1$ to indicate the case that $y_r(t)$ is accessible directly to agent i; otherwise, $\mu_i = 0$.

The *control objective* in this chapter is to determine appropriate triggering condition and effective distributed adaptive controllers $u_i(t)$ for each agent by utilizing continuous local states $(x_i(t))$ and the discrete-time neighbors' states $(x_{j,q}(t_{q,k}^j)$, if $a_{ij} = 1)$ such that:

- All the closed-loop signals are uniformly bounded;
- All the agent outputs can track the desired trajectory $y_r(t)$ as closely as possible.

To achieve the objective, the following assumptions are imposed.

Assumption 7.1.1 *The directed graph $\mathcal{G}$ is balanced and weakly connected. The full knowledge of $y_r(t)$ is directly accessible by at least one agent, i.e., $\sum_{i=1}^{N} \mu_i > 0$.*

Assumption 7.1.2 *The first nth-order derivatives $y_r(t)^{(n)}$ of $y_r(t)$ are bounded, piecewise continuous, and directly known by agent i with $\mu_i = 1$, that is, $|y_r(t)^{(n)}| < F_i$ where F_i is an unknown positive constant.*

7.2 PRELIMINARY CONTROL DESIGN WITH CONTINUOUS COMMUNICATION

In this section, a backstepping-based distributed adaptive consensus control scheme will be presented. In each agent with $\mu_i = 0$, we introduce $\hat{y}_{i,0}(t) \in \Re$ to estimate the unknown reference function $y_r(t)$. The following error variables are defined as

$$z_{i,1} = y_i - \mu_i y_r - (1 - \mu_i)\hat{y}_{i,0} \tag{7.2}$$
$$= \delta_i + (1 - \mu_i)\tilde{y}_{i,0},$$
$$z_{i,q} = x_{i,q} - \alpha_{i,q-1}, \quad q = 2, \ldots, n, \tag{7.3}$$
$$e_i = \sum_{j=1}^{N} a_{ij}(y_i - y_j) + \mu_i(y_i - y_r), \tag{7.4}$$

where $\delta_i = y_i - y_r$ is the tracking error for each agent i. $\tilde{y}_{i,0} = y_r - \hat{y}_{i,0}$ is the estimation error for $\mu_i = 0$.

The virtual control signals $\alpha_{i,q}$ for $q = 1, \ldots, n$ and the actual controller u_i are designed as follows.

$$\alpha_{i,1} = -c_{i,1}z_{i,1} - ke_i + \mu_i \dot{y}_r + (1 - \mu_i)\dot{\hat{y}}_{i,0}, \tag{7.5}$$

$$\alpha_{i,q} = -c_{i,q}z_{i,q} - z_{i,q-1} + \sum_{k=1}^{q-1} \frac{\partial \alpha_{i,q-1}}{\partial x_{i,k}} x_{i,k+1} + \sum_{j=1}^{N} a_{ij} \sum_{k=1}^{q-1} \frac{\partial \alpha_{i,q-1}}{\partial x_{j,k}} x_{j,k+1}$$

$$+ \mu_i \sum_{k=1}^{q} \frac{\partial \alpha_{i,q-1}}{\partial y_r^{(k-1)}} y_r^{(k)} + (1 - \mu_i)\frac{\partial \alpha_{i,q-1}}{\partial \hat{y}_{i,0}} \dot{\hat{y}}_{i,0}, \quad q = 2, \ldots, n, \tag{7.6}$$

$$u_i = \alpha_{i,n} - \psi_i - \varphi_i^T \hat{\theta}_i, \tag{7.7}$$

where $c_{i,1}$, $c_{i,q}$, and k are positive design parameters, $\hat{\theta}_i$ is the estimate of unknown system parameter θ_i.

The parameter update laws $\hat{y}_{i,0}$ and $\hat{\theta}_i$ are designed as

$$\dot{\hat{y}}_{i,0} = -\gamma_{y_{i0}} e_i - \gamma_{y_{i0}} \kappa_{y_{i0}} (\hat{y}_{i,0} - y_{i,0}), \tag{7.8}$$
$$\dot{\hat{\theta}}_i = \Gamma_{\theta_i} \varphi_i z_{i,n} - \Gamma_{\theta_i} \kappa_{\theta_i} (\hat{\theta}_i - \theta_{i,0}), \tag{7.9}$$

where $\gamma_{y_{i0}}$, Γ_{θ_i}, $\kappa_{y_{i0}}$, κ_{θ_i}, $y_{i,0}$, and $\theta_{i,0}$ are positive constants with suitable dimension.

The main results in this section can be formally stated in the following theorem.

Theorem 7.1 *Consider a group of N uncertain agents as modeled in (7.1) with a desired trajectory $y_r(t)$ under Assumptions 7.1.1–7.1.2. By designing the distributed adaptive controllers as (7.7) with parameter update laws (7.8) and (7.9), the following results can be guaranteed.*
(i) All the closed-loop signals are uniformly bounded.
(ii) The tracking error signals $\delta = [\delta_1, \delta_2, \ldots, \delta_N]^T$ will converge to a compact set.
(iii) The upper bound of $\|\delta(t)\|_{[0,T]}^2 = \frac{1}{T} \int_0^T \|\delta(t)\|^2 dt$ can be decreased by choosing suitable design parameters.

Proof. Define a Lyapunov function candidate V_1 as

$$V_1 = \sum_{i=1}^{N} \left[\frac{1}{2}z_{i,1}^2 + \frac{k(1 - \mu_i)}{2\gamma_{y_{i0}}} \tilde{y}_{i,0}^2 \right], \tag{7.10}$$

where $\tilde{y}_{i,0} = y_r - \hat{y}_{i,0}$. Let $\delta = [\delta_1, \ldots, \delta_N]^T$. From (7.2), (7.5), and (7.8), the derivative of V_1 is computed as

$$
\begin{aligned}
\dot{V}_1 \leq & -\frac{k}{2}\delta^T Q\delta + \sum_{i=1}^{N}\left(-c_{i,1}z_{i,1}^2 + z_{i,1}z_{i,2}\right) - \sum_{i=1}^{N}\frac{k(1-\mu_i)\kappa_{y_{i0}}}{2}\tilde{y}_{i,0}^2 \\
& + \sum_{i=1}^{N}\frac{k(1-\mu_i)}{\gamma_{y_{i0}}}|\tilde{y}_{i,0}|F_1 + \sum_{i=1}^{N}\frac{k(1-\mu_i)\kappa_{y_{i0}}}{2}(y_r - y_{i,0})^2 \\
\leq & -\frac{k}{2}\lambda_{\min}(Q)\|\delta\|^2 + \sum_{i=1}^{N}\left(-\frac{1}{2}c_{i,1}z_{i,1}^2 + z_{i,1}z_{i,2}\right) \\
& - \sum_{i=1}^{N}\frac{k(1-\mu_i)\kappa_{y_{i0}}}{4}\tilde{y}_{i,0}^2 + M_1,
\end{aligned}
\tag{7.11}
$$

where $M_1 = \sum_{i=1}^{N}\frac{k(1-\mu_i)}{\gamma_{y_{i0}}^2\kappa_{y_{i0}}}F_1^2 + \sum_{i=1}^{N}\frac{k(1-\mu_i)\kappa_{y_{i0}}}{2}(y_r - y_{i,0})^2$.

The Lyapunov function candidate V_n for the overall system is defined as

$$
V_n = V_1 + \sum_{i=1}^{N}\sum_{q=2}^{n-1}\frac{1}{2}z_{i,q}^2 + \sum_{i=1}^{N}\frac{1}{2}\tilde{\theta}_i^T\Gamma_{\theta_i}^{-1}\tilde{\theta}_i
\tag{7.12}
$$

From (7.2)–(7.9) and (7.11), the derivative of V_n can be computed as

$$
\begin{aligned}
\dot{V}_n \leq & -\frac{k}{2}\lambda_{\min}(Q)\|\delta\|^2 + \sum_{i=1}^{N}\left(-\frac{1}{2}c_{i,1}z_{i,1}^2 - \sum_{q=2}^{n}c_{i,q}z_{i,q}^2\right) \\
& - \sum_{i=1}^{N}\frac{k(1-\mu_i)\kappa_{y_{i0}}}{4}\tilde{y}_{i,0}^2 - \sum_{i=1}^{N}\frac{\kappa_{\theta_i}\|\tilde{\theta}_i\|^2}{2} + M_n \\
\leq & -\frac{k}{2}\lambda_{\min}(Q)\|\delta\|^2 - \sigma V_n + M_n,
\end{aligned}
\tag{7.13}
$$

where $\sigma = \min\{c_{i,1}, 2c_{i,2}, \ldots, 2c_{i,n}, \frac{\gamma_{y_{i0}}\kappa_{y_{i0}}}{2}, \frac{\kappa_{\theta_i}}{\lambda_{\max}(\Gamma_{\theta_i}^{-1})}\}$, $M_n = \sum_{i=1}^{N}\frac{k(1-\mu_i)}{\gamma_{y_{i0}}^2\kappa_{y_{i0}}}F_1^2 +$
$\sum_{i=1}^{N}\frac{k(1-\mu_i)\kappa_{y_{i0}}}{2}(y_r - y_{i,0})^2 + \sum_{i=1}^{N}\frac{\kappa_{\theta_i}\|\theta_i - \theta_{i,0}\|^2}{2}$.

We now establish the results in Theorem 7.1 one by one.

(i) From (7.13), it yields

$$
\dot{V}_n \leq -\sigma V_n + M_n.
\tag{7.14}
$$

By direct integrations of this inequality, we have

$$
V_n(t) \leq V_n(0)e^{-\sigma t} + \frac{M_n}{\sigma}\left(1 - e^{-\sigma t}\right) \leq V_n(0) + \frac{M_n}{\sigma},
\tag{7.15}
$$

which shows that V is uniformly bounded. Thus the error signals $z_{i,q}$ for $1 \leq q \leq n$, $\tilde{\theta}_i$ and $\tilde{y}_{i,0}$ are bounded. From (7.2), δ_i is bounded. Since $\tilde{y}_{i,0} = y_r - \hat{y}_{i,0}$ and y_r is bounded, thus $\hat{y}_{i,0}$ is bounded. From (7.2), (7.5), and (7.6), $x_{i,q}$ and $\alpha_{i,q}$ are bounded. From (7.7), the boundedness of u_i is also ensured. Therefore all the closed-loop signals are uniformly bounded.

(ii) From (7.2), the definitions of V_1 and V_n, we have

$$\|\delta\|^2 \leq \sum_{i=1}^{N} \left[\frac{1}{2} z_{i,1}^2 + \frac{1 - \mu_i}{2} \tilde{y}_{i,0}^2 \right] \leq \xi V_n, \tag{7.16}$$

where $\xi = \max\{1, \frac{\gamma_{y_{10}}}{k}, \ldots, \frac{\gamma_{y_{N0}}}{k}\}$. With (7.15), it further follows that

$$\|\delta(t)\|^2 \leq \xi \left[V_n(0)e^{-\sigma t} + \frac{M_n}{\sigma}(1 - e^{-\sigma t}) \right]. \tag{7.17}$$

This implies that the tracking errors in Euclidean norm will converge to a compact set $E_r = \{\delta \| \|\delta\|^2 \leq \xi(M_n + \varsigma)/\sigma\}$ for $t \geq (1/\sigma) \ln(|V_n(0)\sigma - M_n|/\varsigma)$ with ς an arbitrarily small positive constant.

(iii) From (7.13), we have $\dot{V}_n \leq -\frac{k}{2}\lambda_{\min}(Q)\|\delta\|^2 + M_n$. Integrating both sides of this inequality yields that

$$\begin{aligned}
\|\delta(t)\|_{[0,T]}^2 &= \frac{1}{T} \int_0^T \|\delta(t)\|^2 dt \\
&\leq \frac{2}{k\lambda_{\min}(Q)} \left[\frac{V_n(0) - V_n(T)}{T} + M_n \right] \\
&\leq \frac{2}{k\lambda_{\min}(Q)} \left[\frac{V_n(0)}{T} + M_n \right].
\end{aligned} \tag{7.18}$$

From the definition of V_1, V_n, M_1, M_n, it follows that the upper bound of the overall tracking errors in the mean square sense can be decreased by decreasing κ_{θ_i}, $\kappa_{y_{i0}}$ and increasing k, $c_{i,q}$ for $q = 1, \ldots, n$, Γ_{θ_i}, $\gamma_{y_{i0}}$.

7.3 CONTROL DESIGN WITH EVENT-TRIGGERED COMMUNICATION

In this section, an event-based distributed adaptive control scheme will be presented to achieve the control objective. In addition to Assumptions 7.1.1–7.1.2, the following assumptions are also imposed.

Assumption 7.3.1 $\|\varphi_i(x_i)\| \leq L_{i,1}\|x_i\| + L_{i,0}$, *where* $\|\cdot\|$ *denotes Euclidean norm,* $L_{i,1}$ *and* $L_{i,0}$ *are unknown positive constants.*

Assumption 7.3.2 *The unknown parameter vector* $\theta_i \in \Re^{p_i}$ *is within a compact convex set* C_{θ_i} *with* $\|\theta_i\| \leq L_{\theta_i}$. *The value of* L_{θ_i} *is only known by agent* i.

7.3.1 DESIGN OF TRIGGERING CONDITION

Notations $t_{q,0}^i, t_{q,1}^i, \ldots, t_{q,k}^i, \ldots$ with $0 = t_{q,0}^i < t_{q,1}^i < t_{q,2}^i < \cdots < t_{q,k}^i < t_{q,k+1}^i < \cdots < \infty$, $k \in Z^+$, $i \in \mathcal{V}$, $q = 1, \ldots, n$ are adopted to denote the sequence of event times for agent i to broadcast the information of its qth state to agent j, if $a_{ij} = 1$. $t_{q,0}^i$ is the initial time instant when agent i starts up. For each agent i, the instantaneous information of the qth state for its neighbors' agents is updated only at the time instants $t_{q,k}^j$ for $j \in \mathcal{N}_i$. This indicates that for time $t \in [t_{q,k}^j, t_{q,k+1}^j)$, the neighbor's qth states available for agent i are kept unchanged as $\bar{x}_{j,q}(t) = x_{j,q}(t_{q,k}^j)$, $j \in \mathcal{N}_i$.

The triggering condition for each agent is chosen as

$$t_{q,k+1}^i = \inf\{t > t_{q,k}^i, |x_{i,q}(t) - \bar{x}_{i,q}(t)| > m_q^i\}, \tag{7.19}$$

where $i \in \{0, \mathcal{V}\}$, $\bar{x}_{i,q}(t) = x_{i,q}(t_{q,k}^i)$ and m_q^i is a positive constant to be designed. It is noted from (7.19) that the designed triggering condition for each agent is dependent only on its local state changing rates. Hence, continuous monitoring of neighbors' states can thus be avoided.

7.3.2 DESIGN OF DISTRIBUTED ADAPTIVE CONTROLLERS

The following error variables are defined

$$\bar{z}_{i,1} = y_i - \mu_i y_r - (1 - \mu_i)\hat{\bar{y}}_{i,0} \tag{7.20}$$
$$= \delta_i + (1 - \mu_i)\tilde{\bar{y}}_{i,0},$$
$$\bar{z}_{i,q} = x_{i,q} - \bar{\alpha}_{i,q-1}, \quad q = 2, \ldots, n, \tag{7.21}$$
$$\epsilon_i = \sum_{j=1}^{N} a_{ij}(y_i - \bar{y}_j) + \mu_i(y_i - y_r), \tag{7.22}$$

where $\delta_i = y_i - y_r$ is the tracking error for each agent i and $\tilde{\bar{y}}_{i,0} = y_r - \hat{\bar{y}}_{i,0}$ is the estimation error. $\hat{\bar{y}}_{i,0}$ is the estimate of y_r introduced in agent i with $\mu_i = 0$. Note that the estimator $\hat{y}_{i,0}$ in (7.8) cannot be implemented for the case with event-based communications, as the error variable e_i in (7.4) is unavailable. Define $\bar{x}_{j,q}(t) = x_{j,q}(t_{q,k}^j)$, $j \in \{0, \mathcal{V}\}$, $t \in [t_{q,k}^j, t_{q,k+1}^j)$. $\bar{y}_j(t) = \bar{x}_{j,1}(t)$.

In this case, the virtual control inputs and the actual control input are designed as

$$\bar{\alpha}_{i,1} = -c_{i,1}\bar{z}_{i,1} - k\epsilon_i + \mu_i \dot{y}_r + (1 - \mu_i)\dot{\hat{\bar{y}}}_{i,0}, \tag{7.23}$$

$$\bar{\alpha}_{i,q} = -c_{i,q}\bar{z}_{i,q} - \bar{z}_{i,q-1} + \sum_{k=1}^{q-1} \frac{\partial \alpha_{i,q-1}}{\partial x_{i,k}} x_{i,k+1} + \sum_{j=1}^{N} a_{ij} \sum_{k=1}^{q-1} \frac{\partial \alpha_{i,q-1}}{\partial x_{j,k}} \bar{x}_{j,k+1}$$

$$+ \mu_i \sum_{k=1}^{q} \frac{\partial \alpha_{i,q-1}}{\partial y_r^{(k-1)}} y_r^{(k)} + (1 - \mu_i) \frac{\partial \alpha_{i,q-1}}{\partial \hat{\bar{y}}_{i,0}} \dot{\hat{\bar{y}}}_{i,0}, \quad q = 2, \ldots, n, \tag{7.24}$$

$$u_i = \bar{\alpha}_{i,n} - \psi_i - \varphi_i^T \hat{\theta}_i, \tag{7.25}$$

where $c_{i,1}$, k, and $c_{i,q}$ are positive design parameters. $\hat{\theta}_i$ is the estimate of unknown system parameter θ_i. $\frac{\partial \alpha_{i,q-1}}{\partial x_{i,k}}$, $\frac{\partial \alpha_{i,q-1}}{\partial x_{j,k}}$, $\frac{\partial \alpha_{i,q-1}}{\partial y_r^{(k-1)}}$, and $\frac{\partial \alpha_{i,q-1}}{\partial \hat{y}_{i,0}}$ are the partial derivatives adopted in previously designed $\alpha_{i,q}$ in (7.6).

The estimator $\hat{y}_{i,0}$ and parameter update law for $\hat{\theta}_i$ are designed as

$$\dot{\hat{y}}_{i,0} = -\gamma_{y_{i0}}\epsilon_i - \gamma_{y_{i0}}\kappa_{y_{i0}}(\hat{y}_{i,0} - y_{i,0}), \qquad (7.26)$$

$$\dot{\hat{\theta}}_i = \mathrm{Proj}\{\tau_i\}, \qquad (7.27)$$

where $\tau_i = \Gamma_{\bar{\theta}_i}\varphi_i\bar{z}_{i,n}$, $\mathrm{Proj}\{\cdot\}$ is the projector operator originated from [56]. $\gamma_{y_{i0}}$, $\kappa_{y_{i0}}$, $y_{i,0}$, and $\Gamma_{\bar{\theta}_i}$ are positive constants with appropriate dimension.

Remark 7.1 *By comparing (7.23)–(7.24) and (7.5)–(7.6), it can be seen that $\bar{\alpha}_{i,q}$ is designed in a similar form as $\alpha_{i,q}$. The only difference is that all the continuous neighbors' states $x_{j,q}(t)$ involved in $\alpha_{i,q}$ are replaced with piecewise continuous states $\bar{x}_{j,q}(t)$. The effects due to such replacement will be rigorously analyzed in subsequent section. On the other hand, the partial derivatives terms adopted in $\alpha_{i,q}$ (i.e., $\frac{\partial \alpha_{i,q-1}}{\partial x_{i,k}}$, $\frac{\partial \alpha_{i,q-1}}{\partial x_{j,k}}$, $\frac{\partial \alpha_{i,q-1}}{\partial y_r^{(k-1)}}$, and $\frac{\partial \alpha_{i,q-1}}{\partial \hat{y}_{i,0}}$) are kept unchanged to construct $\bar{\alpha}_{i,q}$. More detailed discussions will be presented in Remark 7.3 and Remark 7.5.*

7.4 STABILITY AND CONSENSUS ANALYSIS

Before providing the main results of this chapter, the following lemmas are introduced for facilitating the stability analysis of the closed-loop system in the sequel.

Lemma 7.1 *By applying the projector operator $\mathrm{Proj}\{\cdot\}$ in [56], the property $-\tilde{\theta}_i^T\Gamma_{\bar{\theta}_i}^{-1}\mathrm{Proj}\{\tau_i\} \leq -\tilde{\theta}_i^T\Gamma_{\bar{\theta}_i}^{-1}\tau_i, \forall\hat{\theta}_i \in \mathcal{C}_{\theta_i}, \theta_i \in \mathcal{C}_{\theta_i}$ exists, where $\tilde{\theta}_i = \theta_i - \hat{\theta}_i$ and $\hat{\theta}_i$ is the estimate of unknown parameter θ_i. $\Gamma_{\bar{\theta}_i}$ is a positive constant matrix with appropriate dimension.*

Lemma 7.2 *The errors between $z_{i,q}$ in (7.3) and $\bar{z}_{i,q}$ in (7.21), $\alpha_{i,q}$ in (7.6) and $\bar{\alpha}_{i,q}$ in (7.24) are bounded. Thus*

$$|z_{i,q} - \bar{z}_{i,q}| \leq \Delta_{z_{i,q}}, \qquad (7.28)$$

$$|\alpha_{i,q} - \bar{\alpha}_{i,q}| \leq \Delta_{\alpha_{i,q}}, \qquad (7.29)$$

where $\Delta_{z_{i,q}}$ and $\Delta_{\alpha_{i,q}}$ are positive constants related to graph parameters Δ_i, μ_i, individual design parameters k, $c_{i,q}$, $\gamma_{y_{i0}}$, $\kappa_{y_{i0}}$, m_q^i and neighbors' design parameters $c_{j,q}$, $\gamma_{y_{j0}}$, $\kappa_{y_{j0}}$, m_q^j for $a_{ij} = 1$ and $q = 1, \ldots, n$.

Proof. Define $s_{y_{i0}} = \hat{y}_{i,0} - \hat{y}_{i,0}$. From (7.8) and (7.26), there is

$$\dot{s}_{y_{i0}} = -\gamma_{y_{i0}}\kappa_{y_{i0}}s_{y_{i0}} + \gamma_{y_{i0}}(e_i - \epsilon_i)$$

$$= -\gamma_{y_{i0}} \kappa_{y_{i0}} s_{y_{i0}} + \gamma_{y_{i0}} \left[\sum_{j=1}^{N} a_{ij} \left(\bar{y}_j - y_j \right) \right]. \qquad (7.30)$$

The solution of this differential equation is computed to satisfy that

$$|s_{y_{i0}}| \le |s_{y_{i0}}(0)| e^{-\gamma_{y_{i0}} \kappa_{y_{i0}} t} + \frac{\Delta_i m}{\kappa_{y_{i0}}} \left(1 - e^{-\gamma_{y_{i0}} \kappa_{y_{i0}} t} \right)$$

$$\le |s_{y_{i0}}(0)| + \frac{\Delta_i m}{\kappa_{y_{i0}}} \triangleq \Delta_{y_{i0}}, \qquad (7.31)$$

where $m = \max\{m_1^1, \dots, m_1^N, \dots, m_n^1, \dots, m_n^N\}$. From (7.2) and (7.20), it further results in

$$|z_{i,1} - \bar{z}_{i,1}| = (1 - \mu_i) \left| \hat{\bar{y}}_{i,0} - \hat{y}_{i,0} \right| \le (1 - \mu_i)|s_{y_{i0}}|$$

$$\le (1 - \mu_i)\Delta_{y_{i0}} \triangleq \Delta_{z_{i,1}}. \qquad (7.32)$$

With (7.4) and (7.22), we have

$$\left| \dot{\hat{y}}_{i,0} - \dot{\hat{\bar{y}}}_{i,0} \right| \le \gamma_{y_{i0}} |\epsilon_i - e_i| + \gamma_{y_{i0}} \kappa_{y_{i0}} |s_{y_{i0}}|$$

$$\le \gamma_{y_{i0}} (\Delta_i m + \kappa_{y_{i0}} \Delta_{y_{i0}}). \qquad (7.33)$$

From (7.5) and (7.23), there is

$$|\alpha_{i,1} - \bar{\alpha}_{i,1}| = \left| -c_{i,1}(z_{i,1} - \bar{z}_{i,1}) - k(e_i - \epsilon_i) + (1 - \mu_i) \left(\dot{\hat{y}}_{i,0} - \dot{\hat{\bar{y}}}_{i,0} \right) \right|$$

$$\le c_{i,1}\Delta_{z_{i,1}} + k\Delta_i m + (1 - \mu_i)\gamma_{y_{i0}}(\Delta_i m + \kappa_{y_{i0}}\Delta_{y_{i0}})$$

$$\triangleq \Delta_{\alpha_{i,1}}. \qquad (7.34)$$

Note that $\alpha_{i,1}$ is the function of $x_{i,1}$, $x_{j,1}$ if $a_{ij} = 1$, y_r, $\dot{y}_r$ if $\mu_i = 1$, $\hat{y}_{i,0}$ if $\mu_i = 0$. It will be shown that all the partial derivatives of $\alpha_{i,1}$ are constants which are associated with the design parameters and the triggering threshold.

$$\frac{\partial e_i}{\partial x_{i,1}} = \Delta_i + \mu_i, \qquad (7.35)$$

$$\frac{\partial e_i}{\partial x_{j,1}} = -a_{ij}, \qquad (7.36)$$

$$\frac{\partial e_i}{\partial y_r} = -\mu_i, \qquad (7.37)$$

$$\frac{\partial \alpha_{i,1}}{\partial x_{i,1}} = -c_{i,1} - k\frac{\partial e_i}{\partial x_{i,1}} + (1 - \mu_i)(-\gamma_{y_{i0}})\frac{\partial e_i}{\partial x_{i,1}}$$

$$= -c_{i,1} - (\Delta_i + \mu_i)[k + \gamma_{y_{i0}}(1 - \mu_i)] \qquad (7.38)$$

$$\frac{\partial \alpha_{i,1}}{\partial x_{j,1}} = -k\frac{\partial e_i}{\partial x_{j,1}} + (1 - \mu_i)(-\gamma_{y_{i0}})\frac{\partial e_i}{\partial x_{j,1}}$$

$$= a_{ij}[k + \gamma_{y_{i0}}(1 - \mu_i)] \tag{7.39}$$

$$\frac{\partial \alpha_{i,1}}{\partial y_r} = -c_{i,1}(-\mu_i) - k\frac{\partial e_i}{\partial y_r} + (1 - \mu_i)(\gamma_{y_{i0}})\frac{\partial e_i}{\partial y_r}$$

$$= (c_{i,1} + k)\mu_i - (1 - \mu_i)\gamma_{y_{i0}}\mu_i \tag{7.40}$$

$$\frac{\partial \alpha_{i,1}}{\partial \dot{y}_r} = \mu_i \tag{7.41}$$

$$\frac{\partial \alpha_{i,1}}{\partial \hat{y}_{i,0}} = c_{i,1}(1 - \mu_i) + (1 - \mu_i)(-\gamma_{y_{i0}}\kappa_{y_{i0}})$$

$$= (1 - \mu_i)(c_{i,1} - \gamma_{y_{i0}}\kappa_{y_{i0}}) \tag{7.42}$$

Therefore, by some straightforward manipulation, we can directly get

$$|z_{i,2} - \bar{z}_{i,2}| \leq |\bar{\alpha}_{i,1} - \alpha_{i,1}| \leq \Delta_{\alpha_{i,1}} \triangleq \Delta_{z_{i,2}}, \tag{7.43}$$

$$|\alpha_{i,2} - \bar{\alpha}_{i,2}| \leq c_{i,2}\Delta_{z_{i,2}} + \Delta_{z_{i,1}} + \sum_{j=1}^{N} a_{ij}\left|\frac{\partial \alpha_{i,1}}{\partial x_{j,1}}\right| m$$

$$+ (1 - \mu_i)\left|\frac{\partial \alpha_{i,1}}{\partial \hat{y}_{i,0}}\right|\gamma_{y_{i0}}(\Delta_i m + \kappa_{y_{i0}}\Delta_{y_{i0}}) \triangleq \Delta_{\alpha_{i,2}}. \tag{7.44}$$

Following the same procedure based on $z_{i,q}$ in (7.3), $\alpha_{i,q}$ in (7.6), $\bar{z}_{i,q}$ in (7.21), $\bar{\alpha}_{i,q}$ in (7.24), we have

$$|z_{i,q} - \bar{z}_{i,q}| \leq |\bar{\alpha}_{i,q-1} - \alpha_{i,q-1}| \leq \Delta_{\alpha_{i,q-1}} \triangleq \Delta_{z_{i,q}}, \tag{7.45}$$

$$|\alpha_{i,q} - \bar{\alpha}_{i,q}| \leq c_{i,q}\Delta_{z_{i,q}} + \Delta_{z_{i,q-1}} + \sum_{j=1}^{N} a_{ij}\sum_{k=1}^{q-1}\left|\frac{\partial \alpha_{i,q-1}}{\partial x_{j,k}}\right| m$$

$$+ (1 - \mu_i)\left|\frac{\alpha_{i,q-1}}{\partial \hat{y}_{i,0}}\right|\gamma_{y_{i0}}\left(\Delta_i m + \kappa_{y_{i0}}\Delta_{y_{i,0}}\right)$$

$$\triangleq \Delta_{\alpha_{i,q}}. \tag{7.46}$$

Remark 7.2 *It can be observed from the Proof of Lemma 7.2 that the partial derivative terms in (7.6) are all constants depending on graph parameters Δ_i, μ_i, individual design parameters k, $c_{i,q}$, $\gamma_{y_{i0}}$, $\kappa_{y_{i0}}$, m_q^i, and neighbors' design parameters $c_{j,q}$, $\gamma_{y_{j0}}$, $\kappa_{y_{j0}}$, m_q^j for $a_{ij} = 1$. This important property enables the utilization of the partial derivatives in designing $\bar{\alpha}_{i,q}$ in (7.24).*

Lemma 7.3 *Define $\bar{z}_1 = [\bar{z}_{1,1}, \bar{z}_{2,1}, \ldots, \bar{z}_{N,1}]^T$, $z_q = [z_{1,q}, z_{2,q}, \ldots, z_{N,q}]^T$ for $2 \leq q \leq n$, $\tilde{\theta} = [\tilde{\theta}_1^T, \ldots, \tilde{\theta}_N^T]^T$ and $\tilde{y}_0$ as the vector of $\tilde{y}_{i,0} = y_r - \hat{y}_{i,0}$ for all the agents with $\mu_i = 0$ and $\tilde{\theta}_i = \theta_i - \hat{\theta}_i$. The states x_i for $1 \leq i \leq N$ satisfy the following inequality*

$$\|x_i\| \leq L_{x_i}\left\|(\bar{z}_1^T, z_2^T, \ldots, z_n^T, \tilde{y}_0^T, \tilde{\theta}^T)^T\right\| + B_{x_i}, \tag{7.47}$$

where L_{x_i} and B_{x_i} are positive constants related to graph parameters and design parameters as stated in Lemma 7.2.

Proof. Note that for the agents with $\mu_i = 0$, we have

$$|\hat{y}_{i0}| = |\hat{\bar{y}}_{i,0} - s_{y_{i0}}| \leq |y_r| + |\tilde{\bar{y}}_{i,0}| + |s_{y_{i0}}| \leq |\tilde{\bar{y}}_{i,0}| + B_{y_{i0}}, \qquad (7.48)$$

where $B_{y_{i0}} = F_0 + \Delta_{y_{i0}}$ and F_0 is the upper bound of $|y_r|$.

From (7.2), we have

$$|x_{i,1}| \leq \sqrt{2}\|(\tilde{z}_{i,1}, \tilde{\bar{y}}_{i,0})^T\| + B_{y_{i0}} + \Delta_{z_{i,1}} \triangleq L_{x_{i1}}\|(\tilde{z}_{i,1}, \tilde{\bar{y}}_{i,0})^T\| + B_{x_{i1}}, \quad (7.49)$$

where $L_{x_{i1}} = \sqrt{2}$ and $B_{x_{i,1}} = B_{y_{i0}} + \Delta_{z_{i,1}}$. From (7.5) and (7.8), there is

$$
\begin{aligned}
|\alpha_{i,1}| &\leq \sqrt{3}\max\{c_{i,1}, k + \gamma_{y_{i0}}, \gamma_{y_{i0}}\kappa_{y_{i0}}\}\|(\tilde{z}_{i,1}, e_i, \tilde{\bar{y}}_{i,0})^T\| \\
&\quad + c_{i,1}\Delta_{z_{i,1}} + \mu_i F_1 + (1 - \mu_i)\gamma_{y_{i0}}\kappa_{y_{i0}}(y_{i,0} + B_{y_{i0}}) \\
&\triangleq L_{\alpha_{i1}}\|(\tilde{z}_{i,1}, e_i, \tilde{\bar{y}}_{i,0})^T\| + B_{\alpha_{i1}},
\end{aligned}
\qquad (7.50)
$$

where $L_{\alpha_{i1}} = \sqrt{3}\max\{c_{i,1}, k + \gamma_{y_{i0}}, \gamma_{y_{i0}}\kappa_{y_{i0}}\}$ and $B_{\alpha_{i1}} = c_{i,1}\Delta_{z_{i,1}} + \mu_i F_1 + (1 - \mu_i)\gamma_{y_{i0}}\kappa_{y_{i0}}(y_{i,0} + B_{y_{i0}})$.

From (7.3), there is

$$
\begin{aligned}
|x_{i,2}| &\leq (1 + L_{\alpha_{i1}})\|(\tilde{z}_{i,1}, z_{i,2}, e_i, \tilde{\bar{y}}_{i,0})^T\| + B_{\alpha_{i1}} \\
&\triangleq L_{x_{i2}}\|(\tilde{z}_{i,1}, z_{i,2}, e_i, \tilde{\bar{y}}_{i,0})^T\| + B_{x_{i2}},
\end{aligned}
\qquad (7.51)
$$

where $L_{x_{i2}} = 1 + L_{\alpha_{i1}}$ and $B_{x_{i2}} = B_{\alpha_{i1}}$. From (7.6), we have

$$
\begin{aligned}
|\alpha_{i,2}| &\leq c_{i,2}|z_{i,2}| + |z_{i,1}| + \left|\frac{\partial \alpha_{i,1}}{\partial x_{i,1}}\right||x_{i,2}| + \sum_{j=1}^{N} a_{ij}\left|\frac{\partial \alpha_{i,1}}{\partial x_{j,1}}\right||x_{j,2}| \\
&\quad + \mu_i \sum_{k=1}^{2}\left|\frac{\partial \alpha_{i,1}}{\partial y_r^{(k-1)}}\right||y_r^{(k)}| \\
&\quad + (1 - \mu_i)\left|\frac{\partial \alpha_{i,1}}{\partial \hat{y}_{i,0}}\right|[\gamma_{y_{i0}}|e_i| + \gamma_{y_{i0}}\kappa_{y_{i0}}(|\hat{y}_{i,0}| + y_{i,0})] \\
&\leq L_{\alpha_{i2}}\|(\tilde{z}_{i,1}, z_{i,2}, e_i, \tilde{\bar{y}}_{i,0})^T\| + B_{\alpha_{i2}} \\
&\quad + \sum_{j=1}^{N} a_{ij}L_{\alpha_{ij2}}\|(\tilde{z}_{j,1}, z_{j,2}, e_j, \tilde{\bar{y}}_{j,0})^T\|,
\end{aligned}
\qquad (7.52)
$$

where $L_{\alpha_{i2}} \triangleq c_{i,2} + 1 + \left|\frac{\partial \alpha_{i,1}}{\partial x_{i,1}}\right|L_{xi2} + (1 - \mu_i)\left|\frac{\partial \alpha_{i,1}}{\partial \hat{y}_{i,0}}\right|\gamma_{y_{i0}} + \left|\frac{\partial \alpha_{i,1}}{\partial \hat{y}_{i,0}}\right|\gamma_{y_{i0}}\kappa_{y_{i0}}$,

$B_{\alpha_{i2}} \triangleq \Delta_{z_{i,1}} + \left|\frac{\partial \alpha_{i,1}}{\partial x_{i,1}}\right|B_{x_{i2}} + \mu_i \sum_{k=1}^{2}\left|\frac{\partial \alpha_{i,1}}{\partial y_r^{(k-1)}}\right|F_k + (1 - \mu_i)\left|\frac{\partial \alpha_{i,1}}{\partial \hat{y}_{i,0}}\right|\gamma_{y_{i0}}\kappa_{y_{i0}}(B_{y_{i0}} + $

$y_{i,0}) + \sum_{j=1}^{N} a_{ij}\left|\frac{\partial \alpha_{i,1}}{\partial x_{j,1}}\right|B_{x_{j2}}$, $L_{\alpha_{ij2}} \triangleq \left|\frac{\partial \alpha_{i,1}}{\partial x_{j,1}}\right|L_{xj2}$.

By following similar analysis, it can be shown that

$$|x_{i,q}| \leq \sum_{i=1}^{N} L_{x_{iq}}\|(\tilde{z}_{i,1}, z_{i,2}, \ldots, z_{i,q}, e_i, \tilde{\bar{y}}_{i,0})^T\| + B_{x_{iq}}$$

$$\leq \left(\sum_{i=1}^{N} L_{x_{iq}}\right) \|(\bar{z}_1^T, z_2^T, \ldots, z_q^T, e^T, \tilde{y}_0^T)^T\| + B_{x_{iq}}, \qquad (7.53)$$

where $e = [e_1, \ldots, e_N]^T$. From (7.4), we have

$$
\begin{aligned}
|e_i| &\leq \sum_{j=1}^{N} a_{ij}(|x_{i,1}| + |x_{j,1}|) + \mu_i(|x_{i,1}| + |y_r|) \\
&\leq (2\sqrt{2}N + \sqrt{2})\|(\bar{z}_1^T, \tilde{y}_0^T)^T\| + \Delta_{e_i} \\
&\triangleq L_{e_i}\|(\bar{z}_1^T, \tilde{y}_0^T)^T\| + \Delta_{e_i}, \qquad (7.54)
\end{aligned}
$$

where $L_{e_i} = 2\sqrt{2}N + \sqrt{2}$ and $\Delta_{e_i} = B_{x_{i,1}} + \sum_{j=1}^{N} a_{ij}(B_{x_{i,1}} + B_{x_{j,1}})$. Thus,

$$
\begin{aligned}
\|e\| &\leq |e_1| + |e_2| + \ldots + |e_n| \\
&\leq NL_{e_i}\|(\bar{z}_1^T, \tilde{y}_0^T)^T\| + N\Delta_{e_i} \\
&\leq NL_{e_i}\|(\bar{z}_1^T, z_2^T, \ldots, z_n^T, \tilde{y}_0^T)^T\| + N\Delta_{e_i}. \qquad (7.55)
\end{aligned}
$$

Using $\left(\sum_{i=1}^{N} L_{x_{iq}}\right) \|(\bar{z}_1^T, z_2^T, \ldots, z_q^T, e^T, \tilde{y}_0^T)^T\| + B_{x_{iq}}$, we can further get

$$
\begin{aligned}
|x_{i,q}| &\leq \left(\sum_{i=1}^{N} L_{x_{iq}}\right) \times [\|(\bar{z}_1^T, z_2^T, \ldots, z_q^T, \tilde{y}_0^T)^T\| + \|e\|] + B_{x_{iq}} \\
&\leq \left(\sum_{i=1}^{N} L_{x_{iq}}\right) \times [(1 + NL_{e_i})\|(\bar{z}_1^T, z_2^T, \ldots, z_n^T, \tilde{y}_0^T)^T\|] \\
&\quad + \left(\sum_{i=1}^{N} L_{x_{iq}}\right) N\Delta_{e_i} + B_{x_{iq}}. \qquad (7.56)
\end{aligned}
$$

Thus,

$$
\begin{aligned}
\|x_i\| &\leq |x_{i,1}| + |x_{i,2}| + \ldots + |x_{i,n}| \\
&\leq \sum_{q=1}^{n}\sum_{i=1}^{N} L_{x_{iq}}[(1 + NL_{e_i})\|(\bar{z}_1^T, z_2^T, \ldots, z_n^T, \tilde{y}_0^T)^T\|] \\
&\quad + \sum_{q=1}^{n}\left(\sum_{i=1}^{N} L_{x_{iq}} N\Delta_{e_i} + B_{x_{iq}}\right) \\
&\triangleq L_{x_i}\|(\bar{z}_1^T, z_2^T, \ldots, z_n^T, \tilde{y}_0^T)^T\| + B_{x_i} \\
&\leq L_{x_i}\|(\bar{z}_1^T, z_2^T, \ldots, z_n^T, \tilde{y}_0^T, \tilde{\theta}^T)^T\| + B_{x_i}, \qquad (7.57)
\end{aligned}
$$

where $L_{x_i} = \sum_{q=1}^{n}\sum_{i=1}^{N} L_{x_{iq}}(1 + NL_{e_i})$ and $B_{x_i} = \sum_{q=1}^{n}\left(\sum_{i=1}^{N} L_{x_{iq}} N\Delta_{e_i} + B_{x_{iq}}\right)$.

The main results in this chapter are formally stated in the following theorem.

Theorem 7.2 *Consider a group of N uncertain agents as modeled in (7.1) with a desired trajectory $y_r(t)$ under Assumptions 7.1.1–7.3.2. By designing the event-triggering communication rules as (7.19) and the distributed adaptive controllers as (7.25), distributed estimators (7.26), parameter update laws (7.27), the following results can be guaranteed.*

(i) All closed-loop signals are uniformly bounded.
(ii) The tracking error signals $\delta = [\delta_1, \delta_2, ..., \delta_N]^T$ will converge to a compact set.
(iii) The upper bound of $\|\delta(t)\|_{[0,T]}^2 = \frac{1}{T}\int_0^T \|\delta(t)\|^2 dt$ can be decreased by choosing suitable design parameters.
(iv) Zeno behavior is excluded.

Proof. Define a Lyapunov function as

$$V_1 = \sum_{i=1}^{N}\left[\frac{1}{2}\bar{z}_{i,1}^2 + \frac{k(1-\mu_i)}{2\gamma_{y_{i0}}}\tilde{\tilde{y}}_{i,0}^2\right]. \tag{7.58}$$

From (7.20), (7.23), and (7.26), the derivative of V_1 is computed as

$$\dot{V}_1 = \sum_{i=1}^{N}\left[-c_{i,1}\bar{z}_{i,1}^2 - k\bar{z}_{i,1}(e_i + \epsilon_i - e_i) + \bar{z}_{i,1}\bar{z}_{i,2}\right]$$

$$+ \sum_{i=1}^{N}\frac{k(1-\mu_i)}{\gamma_{y_{i0}}}\tilde{\tilde{y}}_{i,0}\left(\dot{y}_r - \dot{\tilde{\tilde{y}}}_{i,0}\right)$$

$$\leq \sum_{i=1}^{N}-c_{i,1}\bar{z}_{i,1}^2 - \sum_{i=1}^{N}k\left[\delta_i + (1-\mu_i)\tilde{\tilde{y}}_{i,0}\right]e_i + \sum_{i=1}^{N}\left[k|\bar{z}_{i,1}|\Delta_i m\right.$$

$$\left. + \bar{z}_{i,1}\bar{z}_{i,2} + \frac{k(1-\mu_i)}{\gamma_{y_{i0}}}|\tilde{\tilde{y}}_{i,0}|F_1\right] - \sum_{i=1}^{N}\frac{k(1-\mu_i)}{\gamma_{y_{i0}}}\tilde{\tilde{y}}_{i,0}\dot{\tilde{\tilde{y}}}_{i,0}$$

$$\leq -\frac{k}{2}\lambda_{\min}(Q)\|\delta\|^2 + \sum_{i=1}^{N}\left(-\frac{1}{2}c_{i,1}\bar{z}_{i,1}^2 + \bar{z}_{i,1}\bar{z}_{i,2}\right)$$

$$- \sum_{i=1}^{N}\frac{k(1-\mu_i)\kappa_{y_{i0}}}{4}\tilde{\tilde{y}}_{i,0}^2 + \bar{M}_1, \tag{7.59}$$

where $m = \max\{m_q^1, \ldots, m_q^N\}$ and $\bar{M}_1 = \sum_{i=1}^{N}\frac{k^2\Delta_i^2 m^2}{2c_{i,1}} + \sum_{i=1}^{N}\frac{k(1-\mu_i)\kappa_{y_{i0}}}{2}(y_r - y_{i,0})^2 + \sum_{i=1}^{N}\frac{k(1-\mu_i)}{\gamma_{y_{i0}}^2\kappa_{y_{i0}}}(\gamma_{y_{i0}}\Delta_i m + F_1)^2$.

Define a Lyapunov function candidate V_2 as

$$V_2 = V_1 + \sum_{i=1}^{N}\frac{1}{2}z_{i,2}^2. \tag{7.60}$$

From (7.3) and (7.6), the derivative of V_2 is computed as

$$\dot{V}_2 \le -\frac{k}{2}\lambda_{\min}(Q)\|\delta\|^2 + \sum_{i=1}^{N}\left(-\frac{1}{2}c_{i,1}\bar{z}_{i,1}^2 + \bar{z}_{i,1}\bar{z}_{i,2}\right) - \sum_{i=1}^{N}\frac{k(1-\mu_i)\kappa_{yi0}}{4}\tilde{\bar{y}}_{i,0}^2$$

$$+ \bar{M}_1 + \sum_{i=1}^{N}\left(-c_{i,2}z_{i,2}^2 - z_{i,1}z_{i,2} + z_{i,2}z_{i,3}\right). \tag{7.61}$$

Since $\bar{z}_{i,1}\bar{z}_{i,2} - z_{i,1}z_{i,2} = (\bar{z}_{i,1}-z_{i,1})z_{i,2} + \bar{z}_{i,1}(\bar{z}_{i,2}-z_{i,2}) \le |z_{i,2}|\Delta_{z_{i,1}} + |\bar{z}_{i,1}|\Delta_{z_{i,2}}$, $\dot{V}_2$ can be further derived as

$$\dot{V}_2 \le -\frac{k}{2}\lambda_{\min}(Q)\|\delta\|^2 + \sum_{i=1}^{N}\left(-\frac{1}{4}c_{i,1}\bar{z}_{i,1}^2 - \frac{1}{2}c_{i,2}z_{i,2}^2\right.$$

$$\left. -\frac{k(1-\mu_i)\kappa_{yi0}}{4}\tilde{\bar{y}}_{i,0}^2 + z_{i,2}z_{i,3}\right) + \bar{M}_2, \tag{7.62}$$

where $\bar{M}_2 = \bar{M}_1 + \sum_{i=1}^{N}\left(\frac{1}{c_{i,1}}\Delta_{z_{i,2}}^2 + \frac{1}{2c_{i,2}}\Delta_{z_{i,1}}^2\right)$.

The Lyapunov function for the entire closed-loop system is chosen as

$$V_n = V_2 + \sum_{i=1}^{N}\left(\sum_{i=3}^{n}\frac{1}{2}z_{i,q}^2 + \frac{1}{2}\tilde{\bar{\theta}}_i^T\Gamma_{\theta_i}^{-1}\tilde{\bar{\theta}}_i\right). \tag{7.63}$$

From (7.3), (7.6), (7.25), (7.27), and Lemma 7.1, the derivative of V_n is computed as

$$\dot{V}_n \le -\frac{k}{2}\lambda_{\min}(Q)\|\delta\|^2 + \sum_{i=1}^{N}\left(-\frac{1}{4}c_{i,1}\bar{z}_{i,1}^2 - \frac{1}{2}c_{i,2}z_{i,2}^2 - \sum_{q=3}^{n-1}c_{i,q}z_{i,q}^2\right.$$

$$\left. -\frac{1}{2}c_{i,n}z_{i,n}^2 - \frac{k(1-\mu_i)\kappa_{yi0}}{4}\tilde{\bar{y}}_{i,0}^2\right) - \sum_{i=1}^{N}\frac{\|\tilde{\bar{\theta}}_i\|^2}{2} + \sum_{i=1}^{N}\|\tilde{\bar{\theta}}_i\|\|\varphi_i\|\Delta_{z_{i,n}}$$

$$+ \bar{M}_2 + \sum_{i=1}^{N}\frac{\Delta_{\alpha_{i,n}}^2}{2c_{i,n}} + \sum_{i=1}^{N}\frac{\|\tilde{\bar{\theta}}_i\|^2}{2}. \tag{7.64}$$

According to Assumption 7.3.1, Assumption 7.3.2, and Lemma 7.3, the term $\|\tilde{\bar{\theta}}_i\|\|\varphi_i\|\Delta_{z_{i,n}}$ can be directly derived as

$$\|\tilde{\bar{\theta}}_i\|\|\varphi_i\|\Delta_{z_{i,n}}$$

$$\le L_{\theta_i}(L_{i,1}\|x_i\| + L_{i,0})\Delta_{z_{i,n}}$$

$$\le L_{\theta_i}L_{i,1}\Delta_{z_{i,n}}\left(L_{x_i}\left\|(\bar{z}_1^T, z_2^T, \ldots, z_n^T, \tilde{\bar{y}}_0^T, \tilde{\bar{\theta}}^T)^T\right\| + B_{x_i}\right) + L_{\theta_i}L_{i,0}\Delta_{z_{i,n}}$$

$$\triangleq \frac{c}{2N}\left\|(\bar{z}_1^T, z_2^T, \ldots, z_n^T, \tilde{\bar{y}}_0^T, \tilde{\bar{\theta}}^T)^T\right\|^2 + \bar{M}_{i3}, \tag{7.65}$$

where $c = \min\left\{\frac{1}{4}c_{i,1}, \frac{1}{2}c_{i,2}, c_{i,q}, \frac{k(1-\mu_i)\kappa_{y_{j0}}}{2}, \frac{1}{2}\right\}$ for $q = 3,\ldots,n-1$ and $\mu_i = 0$.

$\bar{M}_{i3} = \frac{N(L_{\theta_i}L_{i,1}\Delta_{z_{i,n}}L_{x_i})^2}{2c} + L_{\theta_i}L_{i,1}\Delta_{z_{i,n}}B_{x_i} + L_{\theta_i}L_{i,0}\Delta_{z_{i,n}}$.

Substituting (7.65) into (7.64) yields that

$$
\begin{aligned}
\dot{V}_n &\leq -\frac{k}{2}\lambda_{\min}(Q)\|\delta\|^2 - c\left\|(\bar{z}_1^T, z_2^T, \ldots, z_n^T, \tilde{y}_0^T, \tilde{\theta}^T)^T\right\|^2 \\
&\quad + \frac{c}{2}\left\|(\bar{z}_1^T, z_2^T, \ldots, z_n^T, \tilde{y}_0^T, \tilde{\theta}^T)^T\right\|^2 + \bar{M}_2 \\
&\quad + \sum_{i=1}^{N}\left(\frac{\Delta_{\alpha_{i,n}}^2}{2c_{i,n}} + \bar{M}_{i3}\right) + \sum_{i=1}^{N}\frac{\|\tilde{\theta}_i\|^2}{2} \\
&\leq -\frac{k}{2}\lambda_{\min}(Q)\|\delta\|^2 - \frac{c}{2}\left\|(\bar{z}_1^T, z_2^T, \ldots, z_n^T, \tilde{y}_0^T, \tilde{\theta}^T)^T\right\|^2 + M^* \\
&\leq -\frac{k}{2}\lambda_{\min}(Q)\|\delta\|^2 - \sigma V_n + M^*,
\end{aligned}
\tag{7.66}
$$

where $M^* = \bar{M}_2 + \sum_{i=1}^{N}\left(\frac{\Delta_{\alpha_{i,n}}^2}{2c_{i,n}} + \bar{M}_{i3}\right) + \sum_{i=1}^{N}\frac{\|\tilde{\theta}_i\|^2}{2}$ and $\sigma = \min\left\{c, \frac{c\kappa_{y_{j0}}}{k(1-\mu_i)}, \frac{c}{\lambda_{\max}(\Gamma_{\bar{\theta}_i}^{-1})}\right\}$.

By following the similar analysis in the proof of Theorem 7.1, the conclusions (i), (ii), and (iii) can be drawn. To avoid repetition, the details are omitted here.

To show the exclusion of Zeno behavior, we shall show that the inter-execution intervals $(t_{q,k+1}^j - t_{q,k}^j)$ for $j \in \mathcal{V}$, $\forall k \in Z^+$ are lower-bounded by a positive constant. Define $\eta_{q,k}^j(t) = x_{j,q}(t) - \bar{x}_{j,q}(t)$ for $t \in [t_{q,k}^j, t_{q,k+1}^j)$, whose derivative is computed as $\frac{d|\eta_{q,k}^j|}{dt} = \frac{d(\eta_{q,k}^j \times \eta_{q,k}^j)^{\frac{1}{2}}}{dt} = \text{sgn}(\eta_{q,k}^j)\dot{\eta}_{q,k}^j \leq \left|\dot{\eta}_{q,k}^j\right|$. Since $\bar{x}_{j,q}(t)$ keeps unchanged for $t \in [t_{q,k}^j, t_{q,k+1}^j)$, we have

$$
\left|\dot{\eta}_{q,k}^j(t)\right| = |x_{j,q+1}|, \quad q = 2,\ldots,n-1,
\tag{7.67}
$$

$$
\left|\dot{\eta}_{n,k}^j(t)\right| = |u_j + \psi_j + \varphi_j^T\theta_j|, \quad \text{for } j \in \mathcal{V}.
\tag{7.68}
$$

From the boundedness of $x_{j,q}$, u_j, φ_j, it is concluded that there exists a positive constant $\iota_{q,j}$ such that $\left|\dot{\eta}_{q,k}^j(t)\right| \leq \iota_q^j$ for $j \in \bar{\mathcal{V}}$. Then the inter-execution intervals must satisfy that $t_{q,k+1}^j - t_{q,k}^j \geq m_q^j/\iota_q^j$, i.e., Zeno behavior is excluded.

Remark 7.3 *Different from Theorem 7.1, the coupling term $\|\tilde{\theta}_i\|\|\varphi_i\|\Delta_{z_{i,n}}$ in (7.64) arises in the derivative of V_n, which is the Lyapunov function defined for the entire closed-loop system in event-based communication case. To effectively handle this term, two additional assumptions, i.e., Assumptions 7.3.1 and 7.3.2 are necessary in this section. Besides, the projector operator Proj{·} in [56] is used to ensure the boundedness of parameter estimate $\hat{\theta}_i$ and the corresponding estimation error $\tilde{\theta}_i$.*

Remark 7.4 *As observed from (7.13), (7.18), and (7.66), the tracking performance of all agent outputs in the mean square error sense is mainly determined by state* *initials, design parameters, and the size of M^*. Besides, it can be seen from $\bar{M}_1$, $\bar{M}_2$, and M^* that event triggering thresholds m_q^j have huge impact on the size of M^*. Clearly, if m_q^j is increased, the triggering times for communication among agents can effectively be reduced. However, the upper bound of $\|\delta\|_{[0,T]}^2$ will be increased with a larger m_q^j. Therefore, determining the values of m_q^j is a tradeoff between the cost of communication resources and consensus tracking performance.*

Remark 7.5 *The main challenge to design backstepping-based distributed adaptive consensus controllers for high-order nonlinear systems with event-based communications lies in the fact that traditional backstepping technique [56] requires differentiating virtual control inputs recursively. If the virtual control input designed in one step involves piecewise continuous signals, computing its derivatives is impossible, thus the subsequent design step is difficult to proceed. To overcome this difficulty, backstepping technique is not adopted directly in designing the virtual controllers and final control law for event-based communication case. Instead, we design the backstepping-based distributed adaptive controllers for all the agents with continuous communications firstly. Thus all the chosen virtual controllers (i.e., $\alpha_{i,q}$) are differentiable. For the case with event-triggered communications, the partial derivative terms $\frac{\partial \alpha_{i,q-1}}{\partial x_{i,k}}$, $\frac{\partial \alpha_{i,q-1}}{\partial x_{j,k}}$, $\frac{\partial \alpha_{i,q-1}}{\partial y_r^{(k-1)}}$, and $\frac{\partial \alpha_{i,q-1}}{\partial \hat{y}_{i,0}}$, which are shown to be constants, are utilized to construct the virtual control input $\bar{\alpha}_{i,q}$ in (7.24). For easier understanding the proposed event-triggered consensus tracking control scheme, its design procedure is provided in Figure 7.1.*

7.5 SIMULATION RESULTS

In this section, we consider a group of 4 pendulum systems [169] modeled with the following dynamics

$$m_i l_i \ddot{\vartheta}_i + m_i g \sin(\vartheta_i) + k_i l_i \dot{\vartheta}_i = u_i, \tag{7.69}$$

where ϑ_i denotes the angle of pendulum, m_i and l_i are the mass $[kg]$ and length of the robe $[m]$, respectively, g denotes the acceleration due to the gravity, k_i is an unknown friction coefficient, and u_i represents an input torque provided by a DC motor. System parameters are chosen the same as these in [169] i.e., $m_i = 1\text{kg}$, $l_i = 1\text{m}$ and $g = 9.8\text{m/s}^2$. The friction coefficients are set as $k_1 = 0.2$, $k_2 = 0.2$, $k_3 = 0.1$, and $k_4 = 0.1$, respectively. The objective is to design distributed adaptive controllers u_i for all the agents such that ϑ_i can track a common desired trajectory $\vartheta_0(t) = 2\cos(0.1t)$ for $1 \le i \le 4$. The communication condition among the 4 agents and $\vartheta_0(t)$ is represented by the directed graph in Figure. 7.2.

Defining the state variables $x_{i,1} = \vartheta_i$ and $x_{i,2} = \dot{\vartheta}_i$ for $i = 1, \ldots, 4$, then (7.69) can be rewritten as the same form as in (7.1).

$$
\begin{aligned}
\dot{x}_{i,1} &= x_{i,2}, \\
\dot{x}_{i,2} &= \frac{1}{m_i l_i} u_i - \frac{g}{l_i} \sin(x_{i,1}) - \frac{k_i}{m_i} x_{i,2}.
\end{aligned} \tag{7.70}
$$

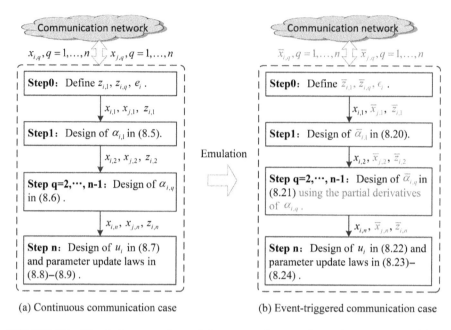

(a) Continuous communication case   (b) Event-triggered communication case

FIGURE 7.1 The design procedure of the proposed event-triggered consensus tracking control scheme.

Clearly, the parameter k_i/m_i is unknown. The triggering condition for inter-agent communication, distributed adaptive controllers and parameter estimators are designed as in (7.19), and (7.25)–(7.27).

In simulation, the state initials including $\vartheta_i(0)$, $\dot{\vartheta}_i(0)$, $\hat{\bar{\theta}}_i(0)$, $\hat{\bar{y}}_{i,0}(0)$ are set as zeros for $i \in \{1, 2, 3, 4\}$. The design parameters are chosen as follows. $k = 0.1$, $c_{i,1} = c_{i,2} = 3$, $\gamma_{\vartheta_i} = 20$, $\Gamma_{\bar{\theta}_i} = 0.3$, $\kappa_{\vartheta_i} = 0.005$, $\vartheta_{i0} = 0.01$, $m_1^i = 0.05$,

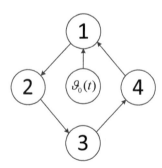

FIGURE 7.2 Information transmission graph for the 4 agents.

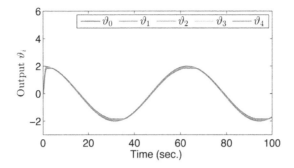

FIGURE 7.3   The outputs ϑ_i, $i = 1, \ldots, 4$.

$m_2^i = 0.05$ for $i = 1, \ldots, 4$. The tracking performance of all the agents' outputs ϑ_i with comparison to ϑ_0, the tracking errors δ_i and states $\dot{\vartheta}_i$ for $i = 1, 2, 3, 4$ are shown in Figures 7.3–7.5, respectively. Figures 7.6–7.7 exhibit the control inputs and the triggering time of all the agents, respectively. The triggering count and minimum inter-event times are provided in Table 7.1. It can be seen that desired tracking performance for all the agents' outputs can be achieved, while all the observed signals are bounded. Moreover, Zeno behavior in each pendulum does not exist. In order to show the effects of triggering thresholds on the tracking performance, m_1^i is changed to $m_1^i = 0.2$ while keeping all the remaining design parameters unchanged. From Figures 7.8–7.9, it can be observed that the tracking performance can be improved by reducing the triggering threshold m_1^i at the cost of increasing the triggering frequency.

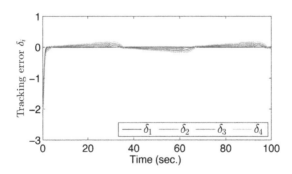

FIGURE 7.4   Tracking errors $\delta_i = \vartheta_i - \vartheta_0$, $i = 1, \ldots, 4$.

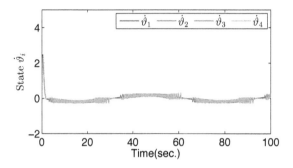

FIGURE 7.5   The states $\dot{\vartheta}_i$, $i = 1, \ldots, 4$.

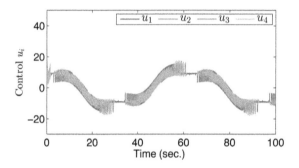

FIGURE 7.6   Control inputs u_i, $i = 1, \ldots, 4$.

7.6   NOTES

In this chapter, the consensus tracking control problem for uncertain high-order nonlinear systems is investigated. Under directed communication condition, two fully distributed adaptive control schemes are proposed with or without event-triggered communication. For the event-triggered case, the continuous monitoring problem of neighbors' states has been removed. The boundedness of all closed-loop signals and the tracking performance have been analyzed. To validate the effectiveness of the proposed control scheme, a numerical case study is provided.

Besides, it should be pointed out that the obtained results formulate a general backstepping-based controller design and stability analysis framework for the chained nonlinear systems to solve the non-differentiability problem of virtual controllers. It can be employed in some other cooperative control problems, such as the quantized consensus for chained nonlinear systems.

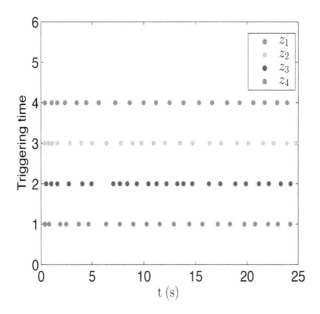

FIGURE 7.7 Triggering times of ϑ_i and $\dot{\vartheta}_i$, $i = 1, \ldots, 4$.

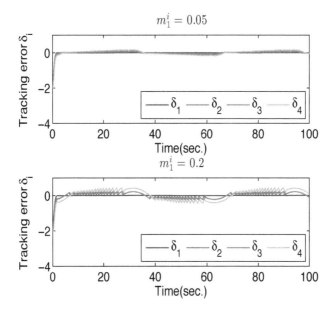

FIGURE 7.8 Errors $\delta_i, i = 1, 2, 3, 4$ for agents with different m_1^i.

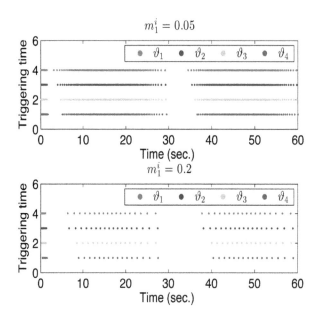

FIGURE 7.9 Triggering times of $\vartheta_i, i = 1, 2, 3, 4$ with different m_1^i.

TABLE 7.1

Event Count and Minimum Inter-Event Times for the States in Each Agent

		Count	Inter-event times (s)
Pendulum 1	ϑ_1	278	0.0203
	$\dot{\vartheta}_1$	122	0.0025
Pendulum 2	ϑ_2	270	0.0209
	$\dot{\vartheta}_2$	981	0.0037
Pendulum 3	ϑ_3	261	0.0215
	$\dot{\vartheta}_3$	1084	0.0042
Pendulum 4	ϑ_4	257	0.0224
	$\dot{\vartheta}_4$	1038	0.0049

8 Distributed Event-Triggered Adaptive Control: Leaderless Case

In Chapter 7, the consensus tracking control problem is solved for high-order nonlinear multi-agent systems with unknown system parameters and event-triggered communication. In this chapter, we aim to discuss the event-triggered leaderless consensus case. It is important to emphasize that the leaderless case is not simply a trivial extension of the leader-follower case. Similar to the leader-following case, the event-triggered leaderless consensus problem is also widely investigated. For linear multi-agent systems, some representative results can be seen in [18, 88, 117, 171]. For low-order nonlinear multi-agent systems, event-based consensus control strategies are presented in [59, 71, 145] under directed or undirected graph condition.

Note that most of the aforementioned consensus algorithms are designed for relatively simple linear or precisely known nonlinear systems. However, system uncertainties inherently exist in real physical systems. Therefore, how to develop distributed event-based consensus strategies for uncertain nonlinear systems is of great significance. To our knowledge, few results are concerned with the distributed adaptive leaderless consensus for uncertain high-order nonlinear systems with event-triggered communication.

In this chapter, we investigate the distributed adaptive leaderless consensus problem for a class of uncertain high-order nonlinear multi-agent systems with directed graph. The communication among the connected agents is based on a pre-designed event condition. To achieve leaderless consensus, a filter is firstly introduced in each agent for system order reduction. Then, an auxiliary system is designed, by using only the latest triggered local and neighboring filtered outputs. Consequently, only one filtered output signal needs be broadcast among the connected agents, which can effectively save communication resources. Besides, to avoid continuously monitoring neighbors' states, the distributed adaptive controllers and event-triggered conditions are co-designed based on a single Lyapunov function. Moreover, an exponentially convergent term is added in the designed triggering condition. Thus asymptotically leaderless consensus can be achieved, while the Zeno behavior in the sense of infinite number of triggers within finite time can be excluded. However, such a triggering condition does not have an explicit lower bound of inter-event times which makes it hard to be implemented in practice. To handle this issue, a switching triggering condition is presented and bounded consensus errors can be shown.

DOI: 10.1201/9781003394372-8

123

8.1 PROBLEM STATEMENT

We consider the following class of uncertain nonlinear multi-agent systems.

$$y_i^{(n)}(t) - \sum_{l=1}^{p_i} \theta_{il} \varphi_{il}(y_i, \dot{y}_i, \ldots, y_i^{(n-1)}) = b_i u_i(t) + d_i(t), \qquad (8.1)$$

where $y_i(t) \in \Re$ is the output of agent i for $i = 1, \ldots, N$. $u_i(t) \in \Re$ is the control input. $b_i \in \Re$, $\theta_{il} \in \Re$ are unknown constant parameters with b_i being nonzero. $\varphi_{il} : \Re^n \to \Re$ is a known nonlinear function. $d_i(t) \in \Re$ is external disturbance.

Let $x_{i,q} = y_i^{(q-1)}$ for $q = 1, \ldots, n$. Nonlinear multi-agent system (8.1) can be rewritten as

$$\dot{x}_{i,q} = x_{i,q+1}, \quad q = 1, \ldots, n-1;$$
$$\dot{x}_{i,n} = b_i u_i(t) + \varphi_i^T \theta_i + d_i(t); \qquad (8.2)$$
$$y_i = x_{i,1}, \qquad (8.3)$$

where $\varphi_i = [\varphi_{i1}, \varphi_{i2}, \ldots, \varphi_{ip_i}]^T$ and $\theta_i = [\theta_{i1}, \theta_{i2}, \ldots, \theta_{ip_i}]^T$.

In this chapter, we aim at designing a distributed adaptive control scheme for a group of agents with event-triggered communication and unknown system parameters such that:

- All the closed-loop signals are globally uniformly bounded;
- All the agents' outputs can achieve a consensus asymptotically, i.e., $\lim_{t\to\infty}[y_i(t) - y_j(t)] = 0, \forall i, j = 1, \ldots, N$.

To achieve this objective, some basic assumptions are needed.

Assumption 8.1.1 *The sign of b_i is known by agent i.*

Assumption 8.1.2 *The graph $\mathcal{G}$ is balanced and weakly connected.*

Assumption 8.1.3 *The disturbance $d_i(t)$ is bounded with an unknown positive upper bound $\bar{d}_i$, i.e., $|d_i(t)| \le \bar{d}_i$*

The following definition and lemmas are stated below, which will be used in the adaptive controller design and stability analysis.

Definition 8.1 ([123, 164]) *The triggering is referred to as Zeno if*

$$\lim_{k\to\infty} t_k = \sum_{k=0}^{\infty}(t_{k+1} - t_k) = t_\infty, \qquad (8.4)$$

where t_∞ is a finite time instant and called Zeno time.

Remark 8.1 *Definition 8.1 is a standard concept initiated from hybrid systems* *[123,164], which characterizes the possibility of infinite number of triggers occurred* *within finite time.*

Lemma 8.1 ([45,73]) *Consider the following system*

$$s(t) = k_1 y(t) + \ldots + k_{n-1} y^{(n-2)}(t) + y^{(n-1)}(t), \tag{8.5}$$

where $k_1, \ldots, k_{n-1}$ are constants and the polynomial $z^{n-1} + k_{n-1} z^{(n-2)} + \ldots + k_1$ is Hurwitz. Then a bounded $s(t)$ results in the bounded $y(t), \ldots, y^{(n-1)}(t)$. Let $\bar{y}(t) = [y(t), \ldots, y^{(n-1)}(t)]^T$. The upper bound of $\bar{y}(t)$ is denoted as $\|\bar{y}(t)\| \leq r e^{-\lambda(t-t_0)} \|\bar{y}(t_0)\| + \frac{r}{\lambda} \sup_{t_0 \leq \tau \leq t} |s(\tau)|$, where $r > 0$ and $\lambda > 0$ are constants related to $k_1, \ldots, k_{n-1}$. Furthermore, $\lim_{t \to \infty} y(t) = 0$ if $\lim_{t \to \infty} s(t) = 0$.

Lemma 8.2 ([172]) *The following property holds*

$$0 \leq |\pi| - \pi \cdot sg(\pi, \eta) \leq \eta,$$

where $\pi \in \Re$ and $sg(\pi, \eta) = \frac{\pi}{\sqrt{\pi^2 + \eta^2}}$. η is an exponentially decaying time-varying function satisfying $\eta > 0$.

8.2 DESIGN OF DISTRIBUTED ADAPTIVE CONTROLLERS

8.2.1 DESIGN OF FILTERS

For each agent i, a filtered output s_i is introduced as follows

$$s_i(t) = k_1 y_i(t) + \ldots + k_{n-1} y_i^{(n-2)}(t) + y_i^{(n-1)}(t), \tag{8.6}$$

where the parameters $k_1, \ldots, k_{n-1}$ are constants chosen to make sure that the polynomial $z^{n-1} + k_{n-1} z^{(n-2)} + \ldots + k_1$ is Hurwitz.

8.2.2 DESIGN OF TRIGGERING CONDITIONS

For agent i, notations $t_0^i, t_1^i, \ldots, t_{k_i}^i, \ldots$ with $0 = t_0^i < t_1^i < \ldots < t_{k_i}^i < \ldots < \infty$, $k_i = 0, 1, \ldots, i \in \mathcal{V}$ are adopted to denote the sequence of triggering time instants to transmit its filtered output $s_i(t_{k_i}^i)$ to its out-neighbors. t_0^i is the initial time instant when a group of agents starts up. For agent i, the instantaneous information of its neighbors' filtered outputs $s_j(t_{k_j}^j)$ is updated only at the time instants $t_{k_j}^j$ for $j \in \mathcal{N}_i$. This implies that the neighbors' filtered outputs available for agent i are kept unchanged during the time intervals $t \in [t_{k_j}^j, t_{k_j+1}^j)$. We denote by $\bar{s}_i(t) = s_i(t_{k_i}^i)$, $t \in [t_{k_i}^i, t_{k_i+1}^i), i \in \mathcal{V}$.

Define a measurement error as

$$\epsilon_i(t) = s_i(t) - \bar{s}_i(t). \tag{8.7}$$

The triggering condition is designed as

$$t^i_{k_i+1} = \inf \left\{ t > t^i_{k_i} \,\middle|\, \epsilon^2_i(t) > \Pi_i(t) \right\},\tag{8.8}$$

where $\Pi_i(t) = \frac{\tau_i}{24\Delta_i} \sum_{j=1}^N a_{ij}[\bar{s}_i(t) - \bar{s}_j(t)]^2 + \frac{1}{3\Delta_i}\varpi_i(t)$ and $\varpi_i(t) = \varsigma_i e^{-\iota_i t}$. $\bar{s}_j(t) = s_j(t^j_{k_j}), t \in [t^j_{k_j}, t^j_{k_j+1}), j \in \mathcal{N}_i$ denotes the latest received neighboring filtered outputs. Δ_i is the diagonal element of in-degree matrix Δ. τ_i, ς_i, and ι_i are positive constants with $0 < \tau_i < 1$.

Remark 8.2 *If $\Pi_i(t)$ in (8.8) is chosen as $\Pi_i(t) = \frac{\tau_i}{24\Delta_i}\sum_{j=1}^N a_{ij}[\bar{s}_i(t) - \bar{s}_j(t)]^2$, there may exist a finite time instant t^* at which $\bar{s}_i(t)$ equals $\bar{s}_j(t)$. Hence, $\Pi_i(t^*) = 0$, which will result in Zeno behavior in agent i at t^*. To solve this problem, inspired by [123, 150], an additional term $\frac{1}{3\Delta_i}\varpi_i(t)$ is added in Π_i to exclude the Zeno behavior as described in Definition 8.1.*

Remark 8.3 *Note that the triggering condition (8.8) depends only on filtered output (i.e., $s_i(t)$ and $s_i(t^i_{k_i})$) and neighbors' latest transmitted filtered outputs (i.e., $s_j(t^j_{k_j})$) if $a_{ij} = 1$. This implies that each agent is not required to monitor the neighboring states continuously to conduct the triggering condition (8.8).*

8.2.3   DESIGN OF DISTRIBUTED ADAPTIVE CONTROLLERS

We design an auxiliary system z_i for each agent, which is governed by

$$\dot{z}_i(t) = -\sum_{j=1}^N a_{ij}[\bar{s}_i(t) - \bar{s}_j(t)].\tag{8.9}$$

Note that the initial condition is chosen as $z_i(0) = x_{i,1}(0)$ to reach a scaling average consensus. For simplicity, time variable t will be omitted subsequently if there is no confusion.

Let

$$\begin{aligned} h_i &= k_1\dot{y}_i + \ldots + k_{n-1}y_i^{(n-1)} \\ &= k_1 x_{i,2} + \ldots + k_{n-1}x_{i,n} \end{aligned}\tag{8.10}$$

and define an error variable as

$$e_i = s_i - z_i.\tag{8.11}$$

Computing the time-derivative of e_i, yields

$$\begin{aligned} \dot{e}_i &= k_1\dot{y}_i + \ldots + k_{n-1}y_i^{(n-1)} + y_i^{(n)} - \dot{z}_i \\ &= b_i u_i + \varphi_i^T \theta_i + d_i + h_i - \dot{z}_i. \end{aligned}\tag{8.12}$$

Therefore, by introducing a filter in each agent, the high-order nonlinear system (8.2) is transformed into a first-order nonlinear system (8.12).

The distributed adaptive controller is designed as

$$u_i = \hat{\varrho}_i \alpha_i, \tag{8.13}$$

$$\alpha_i = -c_i e_i - \varphi_i^T \hat{\theta}_i - h_i - \mathrm{sg}(e_i, \eta_i) \hat{D}_i + 2\dot{z}_i, \tag{8.14}$$

where $\hat{\varrho}_i$ is the estimate of $\frac{1}{b_i}$, $\hat{\theta}_i$ is the estimate of θ_i, $\hat{D}_i$ is the estimate of $bard_i$. $\mathrm{sg}(e_i, \eta_i)$ denotes the function stated in Lemma 8.2, where $\eta_i = \eta_{1i} e^{-\eta_{2i} t}$. c_i, η_{1i} and η_{2i} are positive constants.

The parameter update laws are chosen as

$$\dot{\hat{D}}_i = \gamma_{D_i} e_i \mathrm{sg}(e_i, \eta_i), \tag{8.15}$$

$$\dot{\hat{\varrho}}_i = -\gamma_{\varrho_i} \mathrm{sgn}(b_i) e_i \alpha_i, \tag{8.16}$$

$$\dot{\hat{\theta}}_i = \Gamma_i \varphi_i e_i, \tag{8.17}$$

where γ_{D_i}, γ_{ϱ_i}, and Γ_i are positive design parameters with appropriate dimension.

Remark 8.4 *Note that no knowledge of global parameters is used in the distributed adaptive controller (8.13), the parameter estimators (8.15)–(8.17) and the triggering condition (8.8). Thus the designed adaptive control scheme is totally distributed.*

Remark 8.5 *The block diagram of agent i with the presented adaptive controller and event-based communication is shown in Figure 8.1. It can be observed that only one signal (i.e., $s_i(t_{k_i}^i)$) is required to be transmitted to its neighbors. Compared with some triggering conditions as in [59, 145, 171] with multiple transmitted signals, the presented control algorithm can effectively reduce the communication burden.*

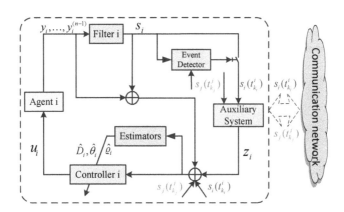

FIGURE 8.1 The block diagram of distributed adaptive control scheme for agent i.

8.3 STABILITY AND CONSENSUS ANALYSIS

We formally state the main results of Section 8.2 in Theorem 8.1.

Theorem 8.1 *Consider a group of N uncertain nonlinear systems (8.1) consisting of the designed triggering condition (8.8), auxiliary systems (8.9), distributed adaptive controllers (8.13), and parameter estimators (8.15)–(8.17). Suppose that Assumptions 8.1.1–8.1.2 hold. It can be concluded that all the closed-loop signals are globally uniformly bounded and all the agents' outputs can reach a consensus asymptotically, i.e., $\lim_{t\to\infty}[y_i(t) - y_j(t)] = 0, \forall i, j = 1, \ldots, N$.*

Proof. A Lyapunov function V_1 is chosen as

$$V_1 = \sum_{i=1}^{N} \frac{1}{2} e_i^2. \tag{8.18}$$

From (8.8), (8.12), (8.13), and (8.14), $\dot{V}_1$ is computed as

$$\dot{V}_1 \leq -\sum_{i=1}^{N} c_i e_i^2 + \sum_{i=1}^{N} e_i[\mathrm{sg}(e_i, \eta_i)\tilde{D}_i - b_i \tilde{\varrho}_i \alpha_i + \varphi_i^T \tilde{\theta}_i]$$
$$+ \sum_{i=1}^{N} e_i \dot{z}_i + \sum_{i=1}^{N} D_i \eta_i, \tag{8.19}$$

where $\tilde{D}_i = D_i - \hat{D}_i$, $\tilde{\varrho}_i = \varrho_i - \hat{\varrho}_i$, and $\tilde{\theta}_i = \theta_i - \hat{\theta}_i$.

The Lyapunov function for the whole closed-loop system is defined as

$$V = V_1 + \sum_{i=1}^{N} \frac{1}{2\gamma_{D_i}} \tilde{D}_i^2 + \sum_{i=1}^{N} \frac{|b_i|}{2\gamma_{\varrho_i}} \tilde{\varrho}_i^2 + \sum_{i=1}^{N} \frac{1}{2}\tilde{\theta}_i^T \Gamma_i^{-1} \tilde{\theta}_i + \sum_{i=1}^{N} \frac{1}{2} z_i^2, \tag{8.20}$$

whose derivative is calculated as

$$\dot{V} = \dot{V}_1 + \sum_{i=1}^{N} \frac{1}{\gamma_{D_i}} \tilde{D}_i(-\dot{\hat{D}}_i) + \sum_{i=1}^{N} \frac{|b_i|}{\gamma_{\varrho_i}} \tilde{\varrho}_i(-\dot{\hat{\varrho}}_i)$$
$$+ \sum_{i=1}^{N} \tilde{\theta}_i^T \Gamma_i^{-1}(-\dot{\hat{\theta}}_i) + \sum_{i=1}^{N} z_i \dot{z}_i. \tag{8.21}$$

From (8.15)–(8.17) and (8.19), (8.21) is further derived as

$$\dot{V} \leq -\sum_{i=1}^{N} c_i e_i^2 - \sum_{i=1}^{N} z_i \sum_{j=1}^{N} a_{ij}(\bar{s}_i - \bar{s}_j) + \sum_{i=1}^{N} e_i \dot{z}_i + \sum_{i=1}^{N} D_i \eta_i$$
$$\leq -\sum_{i=1}^{N} (\epsilon_i + \bar{s}_i - e_i) \sum_{j=1}^{N} a_{ij}(\bar{s}_i - \bar{s}_j) - \sum_{i=1}^{N} c_i e_i^2$$

$$-\sum_{i=1}^{N} e_i \sum_{j=1}^{N} a_{ij}(\bar{s}_i - \bar{s}_j) + \sum_{i=1}^{N} D_i \eta_i$$

$$\leq -\sum_{i=1}^{N} c_i e_i^2 - \sum_{i=1}^{N} \bar{s}_i \sum_{j=1}^{N} a_{ij}(\bar{s}_i - \bar{s}_j)$$

$$+\sum_{i=1}^{N} |\epsilon_i| \sum_{j=1}^{N} a_{ij}|\bar{s}_i - \bar{s}_j| + \sum_{i=1}^{N} D_i \eta_i. \tag{8.22}$$

Based on Assumption 8.1.2, the property $\sum_{i=1}^{N}\sum_{j=1}^{N} a_{ij}\bar{s}_i = \sum_{i=1}^{N}\sum_{j=1}^{N} a_{ij}\bar{s}_j$ holds. (8.22) is further derived as

$$\dot{V} \leq -\sum_{i=1}^{N} c_i e_i^2 - \sum_{i=1}^{N}\sum_{j=1}^{N} a_{ij}(\bar{s}_i^2 - \bar{s}_i\bar{s}_j) + \sum_{i=1}^{N} |\epsilon_i| \sum_{j=1}^{N} a_{ij}|\bar{s}_i - \bar{s}_j| + \sum_{i=1}^{N} D_i \eta_i$$

$$\leq -\sum_{i=1}^{N} c_i e_i^2 - \sum_{i=1}^{N}\sum_{j=1}^{N} a_{ij}(\frac{1}{2}\bar{s}_i^2 - \bar{s}_i\bar{s}_j + \frac{1}{2}\bar{s}_j^2) + \sum_{i=1}^{N} |\epsilon_i| \sum_{j=1}^{N} a_{ij}|\bar{s}_i - \bar{s}_j|$$

$$+\sum_{i=1}^{N} D_i \eta_i$$

$$\leq -\frac{1}{2}\sum_{i=1}^{N}\sum_{j=1}^{N} a_{ij}(\bar{s}_i - \bar{s}_j)^2 - \sum_{i=1}^{N} c_i e_i^2 + \sum_{i=1}^{N} |\epsilon_i| \sum_{j=1}^{N} a_{ij}|\bar{s}_i - \bar{s}_j|$$

$$+\sum_{i=1}^{N} D_i \eta_i. \tag{8.23}$$

By applying Young's inequality $|\epsilon_i||\bar{s}_i - \bar{s}_j| \leq 2\epsilon_i^2 + \frac{(\bar{s}_i - \bar{s}_j)^2}{8}$, (8.23) can be calculated as

$$\dot{V} \leq -\frac{1}{4}\sum_{i=1}^{N}\sum_{j=1}^{N} a_{ij}(\bar{s}_i - \bar{s}_j)^2 - \sum_{i=1}^{N} c_i e_i^2$$

$$+\sum_{i=1}^{N}\sum_{j=1}^{N} a_{ij}\left[2\epsilon_i^2 - \frac{1}{8}(\bar{s}_i - \bar{s}_j)^2\right] + \sum_{i=1}^{N} D_i \eta_i. \tag{8.24}$$

In view of $\sum_{j=1}^{N} a_{ij} = \sum_{j=1}^{N} a_{ji}$, the term $-\frac{1}{4}\sum_{i=1}^{N}\sum_{j=1}^{N} a_{ij}(\bar{s}_i - \bar{s}_j)^2$ can be computed as

$$-\frac{1}{4}\sum_{i=1}^{N}\sum_{j=1}^{N} a_{ij}(\bar{s}_i - \bar{s}_j)^2$$

$$= -\frac{1}{4}\sum_{i=1}^{N}\sum_{j=1}^{N} a_{ij}(s_i - s_j)^2 - \frac{1}{4}\sum_{i=1}^{N}\sum_{j=1}^{N} a_{ij}(\epsilon_i - \epsilon_j)^2$$

$$+ \frac{1}{2}\sum_{i=1}^{N}\sum_{j=1}^{N} a_{ij}|s_i - s_j||\epsilon_i - \epsilon_j|$$

$$\leq -\frac{1}{8}\sum_{i=1}^{N}\sum_{j=1}^{N} a_{ij}(s_i - s_j)^2 + \frac{1}{4}\sum_{i=1}^{N}\sum_{j=1}^{N} a_{ij}(\epsilon_i - \epsilon_j)^2$$

$$\leq -\frac{1}{8}\sum_{i=1}^{N}\sum_{j=1}^{N} a_{ij}(s_i - s_j)^2 + \frac{1}{2}\sum_{i=1}^{N}\sum_{j=1}^{N} a_{ij}(\epsilon_i^2 + \epsilon_j^2)$$

$$\leq -\frac{1}{8}\sum_{i=1}^{N}\sum_{j=1}^{N} a_{ij}(s_i - s_j)^2 + \sum_{i=1}^{N}\sum_{j=1}^{N} a_{ij}\epsilon_i^2. \tag{8.25}$$

Define $s = [s_1^T, \ldots, s_N^T]^T$. Substituting (8.8) and (8.25) into (8.24) yields

$$\dot{V} \leq -\frac{1}{8}\sum_{i=1}^{N}\sum_{j=1}^{N} a_{ij}(s_i - s_j)^2 - \sum_{i=1}^{N} c_i e_i^2 + \sum_{i=1}^{N}(\varpi_i + D_i \eta_i)$$

$$\leq -\frac{1}{8}s^T(\mathcal{L} + \mathcal{L}^T)s - \sum_{i=1}^{N} c_i e_i^2 + \sum_{i=1}^{N}(\varpi_i + D_i \eta_i). \tag{8.26}$$

Integrating both sides of (8.26), yields

$$V(t) + \int_0^t \frac{1}{8}s^T(\tau)(\mathcal{L} + \mathcal{L}^T)s(\tau)d\tau + \sum_{i=1}^{N} c_i \int_0^t e_i(\tau)^2 d\tau$$

$$\leq V(0) + \sum_{i=1}^{N} \bar{\varpi}_i(t) + \sum_{i=1}^{N} D_i \bar{\eta}_i(t), \tag{8.27}$$

where $\bar{\varpi}_i(t) = \int_0^t \varpi_i(\tau)d\tau$ and $\bar{\eta}_i(t) = \int_0^t \eta_i(\tau)d\tau$ are bounded time-varying functions, in which $\varpi_i(t) = \varsigma_i e^{-\iota_i t}$ and $\eta_i(t) = \eta_{1i} e^{-\eta_{2i} t}$ with ς_i, ι_i, η_{1i}, and η_{2i} being positive constants.

We now establish the results in Theorem 8.1 one by one.

1) From (8.27), it can be concluded that e_i, z_i, s_i, $\hat{D}_i$, $\hat{\varrho}_i$, and $\hat{\theta}_i$ are bounded for all agent i. By Lemma 8.1, $y_i, \ldots, y_i^{(n-1)}$ for $i = 1, \ldots, N$ are bounded. Thus $x_{i,1}, \ldots, x_{i,n}$ are bounded. With (8.13) and (8.14), u_i is bounded. Therefore, all the closed-loop signals are globally uniformly bounded.

2) From (8.12), (8.9) and the boundedness of all the closed-loop signals, $\dot{e}_i$ and $\dot{s}_i$ are bounded, i.e., $\dot{e}_i \in L_\infty$ and $\dot{s}_i \in L_\infty$. Thus $s^T(\mathcal{L} + \mathcal{L}^T)\dot{s} \in L_\infty$. On the other hand, it follows from (8.27) that $e_i \in L_2$ and $s^T(\mathcal{L} + \mathcal{L}^T)s \in L_2$. By Barbalat's lemma, it can be concluded that

$$\lim_{t\to\infty} e_i(t) = 0 \text{ and } \lim_{t\to\infty} s^T(t)(\mathcal{L} + \mathcal{L}^T)s(t) = 0. \tag{8.28}$$

Since the graph is a weakly connected and balanced graph, the matrix $\mathcal{L} + \mathcal{L}^T$ is symmetric and positive semi-definite with a eigenvector $\mathbf{1}_N$ associated with eigenvalue 0. Hence, we can further get that

$$\lim_{t\to\infty} [s_i(t) - s_j(t)] = 0. \tag{8.29}$$

Using Lemma 8.1 and (8.6), it follows that

$$
\begin{aligned}
s_i - s_j =& k_1(y_i - y_j) + \ldots + k_{n-1}\left[y_i^{(n-2)} - y_j^{(n-2)}\right] \\
&+ \left[y_i^{(n-1)} - y_j^{(n-1)}\right],
\end{aligned} \tag{8.30}
$$

which indicates that $\lim_{t\to\infty}[y_i(t) - y_j(t)] = 0$. Hence the consensus for all agents is reached asymptotically.

We now start to analyze the final consensus value. Since $\lim_{t\to\infty}[s_i(t) - s_j(t)] = 0$, $\lim_{t\to\infty} e_i(t) = 0$ and $e_i = s_i - z_i$, it follows that $\lim_{t\to\infty}[z_i(t) - z_j(t)] = 0$. Let $z = [z_1, \ldots, z_N]^T$ and $s(t_k) = [s_1(t_{k_1}^1), \ldots, s_N(t_{k_N}^N)]^T$. The auxiliary system (8.9) can be compactly represented as

$$\dot{z} = -\mathcal{L}s(t_k). \tag{8.31}$$

Multiplying both sides of (8.31) by $\mathbf{1}_N = [1, \ldots, 1]^T \in \Re^N$, we obtain that

$$\mathbf{1}_N^T \dot{z} = -\mathbf{1}_N^T \mathcal{L}s(t_k) = 0, \tag{8.32}$$

where the property of the $\mathcal{G}$ is used. From (8.32) and $\lim_{t\to\infty}[z_i(t) - z_j(t)] = 0$, we have

$$\lim_{t\to\infty} z_i(t) = \frac{\sum\limits_{i=1}^{N} z_i(t)}{N} = \frac{\sum\limits_{i=1}^{N} z_i(0)}{N} = \frac{\sum\limits_{i=1}^{N} x_{i,1}(0)}{N}. \tag{8.33}$$

Thus,

$$\lim_{t\to\infty} s_i(t) = \frac{\sum\limits_{i=1}^{N} x_{i,1}(0)}{N}. \tag{8.34}$$

By Lemma 8.1, the system (8.5) is an input to state stable (ISS) linear system. If s_i approaches to a constant as time progresses, then y_i goes to $y_i = \frac{s_i}{k_1}$. Thus the final consensus value is represented as

$$\lim_{t\to\infty} y_i(t) = \lim_{t\to\infty} x_{i,1}(t) = \frac{\sum\limits_{i=1}^{N} x_{i,1}(0)}{Nk_1}. \tag{8.35}$$

Remark 8.6 *The term $2\dot{z}_i$ in (8.14) shows that we design the distributed adaptive controller (8.13) and triggering condition (8.8) simultaneously based on a single closed-loop Lyapunov function (8.20). One $\dot{z}_i$ is used to counteract the term $-\dot{z}_i$ in (8.12). The other $\dot{z}_i$ is utilized to compensate the term $e_i\dot{z}_i$ resulted from the auxiliary system shown in (8.22) when the triggering condition is designed.*

Remark 8.7 *From (8.35), we can see that a scaling average consensus is reached with a scaling factor k_1. We can tune k_1 to change the final consensus value if it is necessary. As a special case, a standard average consensus is achieved if k_1 is set as 1.*

Theorem 8.2 *With the same conditions as stated in Theorem 8.1, Zeno behavior in the sense of Definition 8.1 is ruled out in each agent i.*

Proof. Assume that the triggering condition (8.8) in agent i happens at time $t_{k_i}^i$. Obviously, $\epsilon_i(t_{k_i}^i) = 0$. During the time interval $[t_{k_i}^i, t_{k_i+1}^i)$, the derivative of ϵ_i is calculated as

$$\dot{\epsilon}_i = \dot{s}_i = b_i u_i + \varphi_i^T \theta_i + d_i + h_i. \tag{8.36}$$

Then,

$$\epsilon_i = \int_{t_{k_i}^i}^t \dot{s}_i(s) ds = \int_{t_{k_i}^i}^t (b_i u_i + \varphi_i^T \theta_i + d_i + h_i) ds. \tag{8.37}$$

Since all the closed-loop signals are bounded as shown in Theorem 8.1 and $|d_i|$ is upper bounded by $\bar{d}_i$, there exists a positive constant $\bar{\varsigma}_i$ such that $|b_i u_i + \varphi_i^T \theta_i + d_i + h_i| \leq \bar{\varsigma}_i$. From (8.37), we have $\epsilon_i \leq (t - t_{k_i}^i)\bar{\varsigma}_i$. From (8.8), we can see that the next event will not be triggered before $\epsilon_i^2 = \frac{\tau_i}{24\Delta_i} \sum_{j=1}^N a_{ij}(\bar{s}_i - \bar{s}_j)^2 + \frac{1}{3\Delta_i}\varpi_i(t)$. Defining the inter-event time interval between two successive events as

$$\tau_{k_i}^i = t_{k_i+1}^i - t_{k_i}^i, k_i = 0, 1, \ldots \tag{8.38}$$

Then the inter-event time interval can be computed as

$$\tau_{k_i}^i = \frac{\sqrt{\frac{\tau_i}{24\Delta_i} \sum_{j=1}^N a_{ij}(\bar{s}_i - \bar{s}_j)^2 + \frac{1}{3\Delta_i}\varpi_i(t_{k_i}^i + \tau_{k_i}^i)}}{\bar{\varsigma}_i}. \tag{8.39}$$

Since $\bar{\varsigma}_i$ is positive, bounded and $\varpi_i(t) > 0$, $\tau_{k_i}^i$ is positive. Moreover, $\tau_{k_i}^i$ exists in any finite time and strictly depends on the time variable $t_{k_i}^i$. Note that the numerator term in (8.39) approaches to 0 only when $\varpi_i(t)$ approaches to 0 as t goes to ∞, which implies that $t_{k_i}^i$ approaches to ∞. Therefore, Zeno behavior, as stated in Definition 8.1, is ruled out in each agent i.

Remark 8.8 *We now start to analyze the effects of the exponentially convergent term $\varpi_i(t)$ in (8.8) on the entire closed-loop system. Intuitively, the bigger the amplitude of $\varpi_i(t)$ is, the bigger the triggering threshold is such that the inter-event times will become large. On the other hand, from (8.26), it can be seen that $\varpi_i(t)$ will affect the system convergence rate. We define $\varpi_{i1}(t) = \varsigma_{i1} e^{-\iota_{i1} t}$ and $\varpi_{i2}(t) = \varsigma_{i2} e^{-\iota_{i2} t}$ with $0 < \varsigma_{i1} < \varsigma_{i2}$ and $0 < \iota_{i2} < \iota_{i1}$ such that $0 < \varpi_{i1}(t) < \varpi_{i2}(t)$. From (8.26), we have $\left|\dot{V}_{\varpi_{i1}}\right| > \left|\dot{V}_{\varpi_{i2}}\right|$. This indicates that the convergence rate of $V_{\varpi_{i1}}$ is greater than that of $V_{\varpi_{i2}}$. Therefore, it is a tradeoff between the communication burden and the convergence rate to determine the form of $\varpi_i(t)$.*

Remark 8.9 *The main difficulties in this chapter are stated as follows. i) How to design distributed controllers for uncertain high-order nonlinear systems with directed communication graph, by using only discrete neighboring states; ii) How to design triggering conditions while avoiding continuous monitoring of neighboring states. To overcome these difficulties, a filter is first introduced to reduce the order of the systems. Then, to reach a leaderless consensus under connected and balanced graph condition, an auxiliary system is introduced, by using the latest triggered local and neighboring filtered outputs. Finally, the distributed adaptive controllers and triggering conditions are designed simultaneously based on a single closed-loop Lyapunov function. As a result, continuous monitoring of neighboring states as required in many existing literatures can be avoided.*

8.4 A CONSENSUS PERFORMANCE-RELATED SWITCHING TRIGGERING STRATEGY

In subsection 8.2.3, a totally distributed adaptive consensus control scheme is presented. All the agents' outputs can achieve a consensus asymptotically while Zeno behavior as stated in Definition 8.1 is excluded. However, some limitations still exist in the designed triggering condition (8.8). 1) No explicit lower bound of the inter-event intervals can be derived. As pointed in [89], such a triggering condition is hard to be implemented in practice. 2) As $\varpi_i(t)$ decays, the triggering condition (8.8) is easily triggered due to the existence of external disturbances. 3) As analyzed in Remark 8.8, the additional term $\varpi_i(t)$ in (8.8) will decrease the consensus convergence rate. To remove these limitations, a switching triggering strategy is presented in this section.

The switching triggering condition is designed as

$$t_{k_i+1}^i = \inf \left\{ t > t_{k_i}^i \,\middle|\, \epsilon_i^2(t) > \Pi_i(t) \right\} \tag{8.40}$$

with

$$\Pi_i(t) = \begin{cases} \zeta_i(t), & \zeta_i(t) \geq M_i \\ m_i, & \zeta_i(t) < M_i \end{cases}, \tag{8.41}$$

where $\zeta_i(t) = \frac{\tau_i}{24\Delta_i} \sum_{j=1}^{N} a_{ij} [\bar{s}_i(t) - \bar{s}_j(t)]^2$ is a piecewise continuous function which jumps at time instants $t_{k_i}^i$ and $t_{k_j}^j$. M_i is a positive switching parameter. m_i is a positive triggering threshold and satisfies $0 < 2m_i < M_i$. τ_i and Δ_i are the same as those stated in Section 8.2.3.

Remark 8.10 *Compared with (8.8), the additional term $\varpi_i(t)$ in $\Pi_i(t)$ is removed and a switching triggering mechanism is introduced instead. When $\zeta_i(t)$ is greater than the switching value M_i, a state-dependent triggering threshold is adopted. As analyzed in Remark 8.8, the consensus convergence rate can be increased by removing the additional term $\varpi_i(t)$. When $\zeta_i(t)$ is less than the switching value M_i, the state-dependent triggering threshold switches to a constant triggering threshold m_i. The main advantages of a constant triggering threshold lie in two aspects: 1) A constant threshold is robust to external disturbances; 2) Zeno triggering can be*

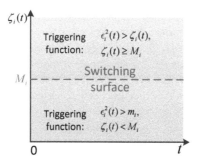

FIGURE 8.2   The switching triggering mechanism in (8.40).

easily excluded and explicit lower bound of the inter-event intervals exists such that the designed triggering mechanism is easier to be implemented. The switching triggering mechanism in (8.40) is illustrated in Figure 8.2.

The following theorem shows that by replacing the triggering condition (8.8) by the switching one (8.40), bounded consensus can be ensured.

Theorem 8.3 *Consider N uncertain nonlinear systems (8.1) under Assumptions 8.1.1–8.1.2. By applying the triggering condition (8.40), auxiliary systems (8.9), distributed adaptive controllers (8.13), and parameter estimators (8.15)–(8.17), all the closed-loop signals are globally uniformly bounded and all the agents' outputs can reach a bounded consensus, i.e., $\lim_{t\to\infty}[y_i(t) - y_j(t)] \le \bar{\varepsilon}_i, \forall i, j = 1, \ldots, N,$ where $\bar{\varepsilon}_i$ is a positive constant.*

Proof. Define the same Lyapunov function V as that in (8.20). With the designed auxiliary systems (8.9), distributed adaptive controllers (8.13), and parameter estimators (8.15)–(8.17), $\dot{V}$ is computed as

$$\dot{V} \le -\frac{1}{8}\sum_{i=1}^{N}\sum_{j=1}^{N}a_{ij}(s_i - s_j)^2 - \sum_{i=1}^{N}c_i e_i^2$$
$$+ \sum_{i=1}^{N}\sum_{j=1}^{N}a_{ij}\left[3\epsilon_i^2 - \frac{1}{8}(\bar{s}_i - \bar{s}_j)^2\right] + \sum_{i=1}^{N}D_i\eta_i. \qquad (8.42)$$

Since the designed switching triggering condition (8.40) has two kinds of triggering thresholds (i.e., $\zeta_i(t)$ and m_i), there are two cases needed to be discussed.

We first consider the case with state-dependent triggering threshold $\zeta_i(t)$, i.e., $\zeta_i(t) \ge M_i$. From the triggering condition (8.40), $\dot{V}$ is further calculated as

$$\dot{V} \le -\frac{1}{8}\sum_{i=1}^{N}\sum_{j=1}^{N}a_{ij}(s_i - s_j)^2 - \sum_{i=1}^{N}c_i e_i^2 + \sum_{i=1}^{N}D_i\eta_i. \qquad (8.43)$$

Following similar analysis of the derivative of the Lyapunov function V in (8.26), it follows from (8.43) that $\lim_{t\to\infty}[s_i(t) - s_j(t)] = 0$.

With $\epsilon_i = s_i - \bar{s}_i$ and $\sum_{j=1}^N a_{ij} = \sum_{j=1}^N a_{ji}$, the following inequality holds.

$$\sum_{i=1}^N \sum_{j=1}^N a_{ij}(\bar{s}_i - \bar{s}_j)^2 \leq 2 \sum_{i=1}^N \sum_{j=1}^N a_{ij}(s_i - s_j)^2 + 2 \sum_{i=1}^N \sum_{j=1}^N a_{ij}(\epsilon_i - \epsilon_j)^2$$

$$\leq 2 \sum_{i=1}^N \sum_{j=1}^N a_{ij}(s_i - s_j)^2 + 8 \sum_{i=1}^N \sum_{j=1}^N a_{ij}\epsilon_i^2, \tag{8.44}$$

which implies $\sum_{j=1}^N a_{ij}(\bar{s}_i - \bar{s}_j)^2 \leq 2\sum_{j=1}^N a_{ij}(s_i - s_j)^2 + 8\Delta_i\epsilon_i^2$.

From the definition of the triggering condition in (8.40), we can further get

$$3\Delta_i\epsilon_i^2 \leq \frac{\tau_i}{8} \sum_{j=1}^N a_{ij}(\bar{s}_i - \bar{s}_j)^2 \leq \frac{\tau_i}{4} \sum_{j=1}^N a_{ij}(s_i - s_j)^2 + \Delta_i\epsilon_i^2, \tag{8.45}$$

which implies $2\Delta_i\epsilon_i^2 \leq \frac{\tau_i}{4} \sum_{j=1}^N a_{ij}(s_i - s_j)^2$.

Since $\lim_{t\to\infty}[s_i(t) - s_j(t)] = 0$, it follows that $\lim_{t\to\infty} \epsilon_i(t) = 0$. Recalling (8.44), it can be concluded that $\lim_{t\to\infty} \sum_{j=1}^N a_{ij}[\bar{s}_i(t) - \bar{s}_j(t)]^2 = 0$, which implies $\lim_{t\to\infty} \zeta_i(t) = 0$. Therefore, there must exist a finite time instant t_{i1}^* such that $\zeta_i(t) < M_i$ for $t > t_{i1}^*$, if $\zeta_i(t_0) \geq M_i$ at the initial time instant t_0. Then the triggering threshold will switch to a constant threshold.

We then discuss the case with constant triggering threshold m_i, i.e., $\zeta_i(t) < M_i$. $\dot{V}$ is calculated as

$$\dot{V} \leq -\frac{1}{8} \sum_{i=1}^N \sum_{j=1}^N a_{ij}(s_i - s_j)^2 - \sum_{i=1}^N c_i e_i^2$$

$$+ \sum_{i=1}^N \sum_{j=1}^N a_{ij}\left[3m_i - \frac{1}{8}(\bar{s}_i - \bar{s}_j)^2\right] + \sum_{i=1}^N D_i\eta_i$$

$$\leq -\frac{1}{8} \sum_{i=1}^N \sum_{j=1}^N a_{ij}(s_i - s_j)^2 - \sum_{i=1}^N c_i e_i^2$$

$$+ \sum_{i=1}^N \left[3\Delta_i m_i - \frac{1}{8} \sum_{j=1}^N a_{ij}(\bar{s}_i - \bar{s}_j)^2\right] + \sum_{i=1}^N D_i\eta_i. \tag{8.46}$$

When $\frac{1}{16}\sum_{j=1}^N a_{ij}(\bar{s}_i - \bar{s}_j)^2 \geq 3\Delta_i m_i$, (8.46) is further calculated as

$$\dot{V} \leq -\frac{1}{8} \sum_{i=1}^N \sum_{j=1}^N a_{ij}(s_i - s_j)^2 - \sum_{i=1}^N c_i e_i^2$$

$$-\frac{1}{16}\sum_{i=1}^{N}\sum_{j=1}^{N}a_{ij}(\bar{s}_i - \bar{s}_j)^2 + \sum_{i=1}^{N}D_i\eta_i$$

$$\leq -\frac{1}{8}\sum_{i=1}^{N}\sum_{j=1}^{N}a_{ij}(s_i - s_j)^2 - \sum_{i=1}^{N}c_ie_i^2 + \sum_{i=1}^{N}D_i\eta_i. \qquad (8.47)$$

Similar to the analysis for inequalities (8.26) and (8.43), we can conclude that there must exist a finite time instant t_{i2}^* such that $\zeta_i(t) < 2m_i$ for $t > t_{i2}^*$, if $\zeta_i(t_0) \geq 2m_i$. That is, $\zeta_i(t)$ will converge to the region $0 < \zeta_i(t) \leq 2m_i$ as time progresses. Recalling $0 < 2m_i < M_i$, it implies that the triggering condition (8.40) will no longer switch from the constant threshold m_i to the state-dependent threshold $\zeta_i(t)$.

Integrating both sides of (8.47) gets

$$V(t) + \sum_{i=1}^{N}c_i\int_0^t e_i(\tau)^2 d\tau + \frac{1}{8}\sum_{i=1}^{N}\sum_{j=1}^{N}a_{ij}\int_0^t [s_i(\tau) - s_j(\tau)]^2 d\tau$$

$$\leq V(0) + \sum_{i=1}^{N}D_i\bar{\eta}_i. \qquad (8.48)$$

It follows from (8.48) that e_i, z_i, s_i, $\hat{D}_i$, $\hat{\varrho}_i$, and $\hat{\theta}_i$ are bounded for all agent i. Following the similar analysis in Theorem 8.1, all the closed-loop signals are globally uniformly bounded.

Next, we discuss the final consensus error. In the previous discussion, we know that $\lim_{t\to\infty}\zeta_i(t) = \lim_{t\to\infty}\frac{\tau_i}{24\Delta_i}\sum_{j=1}^{N}a_{ij}[\bar{s}_i(t) - \bar{s}_j(t)]^2 \leq 2m_i$, which implies $\lim_{t\to\infty}a_{ij}|\bar{s}_i(t) - \bar{s}_j(t)| \leq \lim_{t\to\infty}\sum_{j=1}^{N}a_{ij}|\bar{s}_i(t) - \bar{s}_j(t)| < \sqrt{\frac{48\Delta_i m_i}{\tau_i}}$. Using the triggering condition (8.40) for $\zeta_i(t) < M_i$, we have $|\epsilon_i(t)| \leq \sqrt{m_i}$. Then, it can be further derived that

$$\lim_{t\to\infty}a_{ij}|s_i(t) - s_j(t)| \leq \lim_{t\to\infty}a_{ij}|\bar{s}_i(t) - \bar{s}_j(t)| + |\epsilon_i(t)| + |\epsilon_j(t)|$$

$$\leq \sqrt{\frac{48\Delta_i m_i}{\tau_i}} + \sqrt{m_i} + \sqrt{m_j}, \qquad (8.49)$$

which indicates that for any agent i and agent j, there exists a positive constant Δ_{s_i} such that $\lim_{t\to\infty}[s_i(t) - s_j(t)] \leq \Delta_{s_i}$, $i, j \in \mathcal{V}$.

Define $y_{ij}^{(l)}(t) = y_i^{(l)}(t) - y_j^{(l)}(t)$ for $l = 0, 1, \ldots, n-1$ and $s_{ij}(t) = s_i(t) - s_j(t)$. Let $\bar{y}_{ij}(t) = [y_{ij}(t), \ldots, y_{ij}^{(l)}(t)]^T$. By Lemma 8.1 and (8.30), the upper bound of $\bar{y}_{ij}(t)$ is written as

$$\|\bar{y}_{ij}(t)\| \leq re^{-\lambda(t-t_0)}\|\bar{y}_{ij}(t_0)\| + \frac{r}{\lambda}\sup_{t_0\leq\tau\leq t}|s_{ij}(\tau)|, \qquad (8.50)$$

where $r > 0$ and $\lambda > 0$ are constants related to $k_1, \ldots, k_{n-1}$.

From (8.50) and $\lim_{t\to\infty}[s_i(t) - s_j(t)] \leq \Delta_{s_i}$, we can further have

$$\lim_{t\to\infty}[y_i(t) - y_j(t)] \leq \lim_{t\to\infty}\|\bar{y}_{ij}(t)\| \leq \frac{r}{\lambda}\Delta_{s_i} \triangleq \bar{\varepsilon}_i. \qquad (8.51)$$

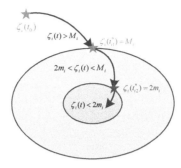

FIGURE 8.3 The schematic diagram of the evolution of $\zeta_i(t)$ as time progresses.

Therefore, all the agents' outputs can achieve a bounded consensus as time progresses.

Moreover, as done in the proof of Theorem 8.1, we can derive that the final consensus value will vary around the scaling average value of $x_{i,1}(0)$, that is,

$\lim_{t\to\infty} y_i(t) = \lim_{t\to\infty} x_{i,1}(t) = \frac{\sum_{i=1}^{N} x_{i,1}(0)}{Nk_1} + \delta_i(t)$, where $\delta_i(t)$ is a time-varying bounded consensus error.

Remark 8.11 *For easier understanding the switching triggering mechanism (8.40), the evolution of $\zeta_i(t)$ is provided in Figure 8.3. According to the proof of Theorem 8.3, $\zeta_i(t)$ will decay to M_i and cross it if $\zeta_i(t) \geq M_i$ initially. Then the state-dependent triggering threshold $\zeta_i(t)$ will be changed to a constant triggering threshold m_i. Furthermore, $\zeta_i(t)$ will eventually converge to $0 < \zeta_i(t) < 2m_i$ as shown in Figure 8.3. Note that the size of $\zeta_i(t)$ is closely dependent on the triggering parameters τ_i and m_i.*

Remark 8.12 *It can be seen from (8.49) and (8.51) that the consensus error is mainly determined by Δ_{s_i}, which is closely related to the triggering threshold m_i. Inequality (8.49) indicates that the consensus error can be reduced by decreasing m_i. However, this will result in decreased inter-event time intervals, which is unsatisfactory in practical implementation. Therefore, the selection of a suitable m_i is a trade-off issue between the consensus error and reduced communication cost.*

Theorem 8.4 *With the same conditions as Theorem 8.3, Zeno triggering does not exist in each agent i. Moreover, the inter-event time internals (i.e., $t^i_{k_i+1} - t^i_{k_i}, k_i = 0, 1, \ldots$) have a positive constant lower bound.*

Proof. As done in the proof of Theorem 8.2, we define the inter-event time as $\tau^i_{k_i} = t^i_{k_i+1} - t^i_{k_i}, k_i = 0, 1, \ldots$. Then we can also have $\epsilon_i \leq \tau^i_{k_i} \bar{\varsigma}_i$, where $\bar{\varsigma}_i$ is a positive constant. The triggering condition (8.40) shows that the next event is not triggered when $\epsilon^2_i = m_i$. Hence the lower bound of the inter-event intervals is derived as $\tau^i_{k_i} \geq \frac{\sqrt{m_i}}{\bar{\varsigma}_i}$. Therefore, Zeno triggering is avoided in agent i.

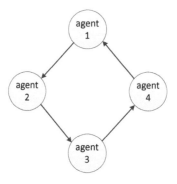

FIGURE 8.4   Communication graph among the 4 pendulums.

8.5   SIMULATION RESULTS

In this section, 4 pendulum systems [169] as follows are utilized to illustrate our proposed control scheme

$$\dot{\vartheta}_i = \omega_i;$$
$$J_i \dot{\omega}_i = u_i - M_i g l_i \sin(\vartheta_i) - f_{di}\omega_i + d_i(t); \ i = 1,\dots 4. \qquad (8.52)$$

where ϑ_i is the angle and ω_i is the angular velocity. $M_i = 1/3\text{kg}$ is the mass. $l_i = 2/3\text{m}$ denotes the length. $g = 9.8\text{m/s}^2$ is the acceleration of gravity. $f_{di} = 0.2$, $J_i = 4/3 M_i l_i^2$ are the frictional factor and the rotary inertia, respectively. $d_i(t)$ denotes unknown external disturbances. M_i, l_i, J_i, g, and f_{di} are unknown system parameters. Let $b_i = J_i^{-1}$, $\theta_i = [J_i^{-1} M_i g l_i, J_i^{-1} f_{di}]^T$ and $\varphi_i = [-\sin(\vartheta_i), -\omega_i]^T$. The communication graph is provided in Figure 8.4.

In simulation, the initials including $\omega_1(0)$, $\omega_2(0)$, $\omega_3(0)$, $\omega_4(0)$, $\hat{D}_i(0)$, $\hat{\varrho}_i(0)$, $\hat{\theta}_i(0)$ for $i = 1,\dots,4$ are set as zeros. $\vartheta_1(0) = z_1(0) = 2$, $\vartheta_2(0) = z_2(0) = 1.5$, $\vartheta_3(0) = z_3(0) = 1$, $\vartheta_4(0) = z_4(0) = 2.5$. $d_1(t) = 0.05\sin(0.1t)$, $d_2(t) = 0.02\sin(0.1t)$, $d_3(t) = 0.05\sin(0.1t)$ and $d_4(t) = 0.02\sin(0.1t)$. The design parameters for the distributed adaptive controller and parameter estimators are chosen as $k_1 = 1$, $c_i = 3$ $\gamma_{c_i} = \gamma_{D_i} = \gamma_{\varrho_i} = 1$, $\Gamma_i = \text{diag}\{1,1\}$, and $\eta_i = 0.02e^{0.03t}$. Besides, design parameters for triggering conditions are chosen as follows.

- Triggering condition (8.8): $\tau_i = \frac{1}{2}$, $\varsigma_i = 0.01$, and $\iota_i = 0.6$.
- Triggering condition (8.40): $\tau_i = \frac{1}{2}$, $M_i = 0.2$, and $m_i = 0.01$.

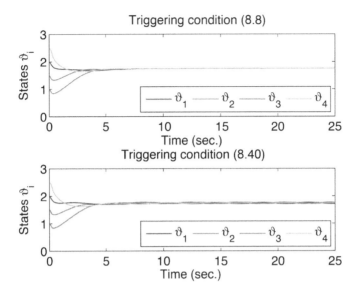

FIGURE 8.5 The states ϑ_i, $i = 1, \ldots, 4$.

The trajectories of angles and angular velocities for each agent are provided in Figures 8.5 and 8.6. Clearly, the observed signals are bounded, and asymptotical and bounded consensus can be reached, by adopting the triggering conditions (8.8) and (8.40), respectively. Figure 8.7 exhibits the performance of corresponding control inputs. The triggering times, triggering counts, and inter-event times of signal transmission are shown in Figures 8.8–8.9 and Tables 8.1–8.2. From Figure 8.8, we can see that as $\varpi_i(t)$ decays, the triggering frequency of the triggering condition (8.8) will become fast for the case with external disturbances. Figure 8.9 shows that the switched triggering condition (8.40) is more robust to external disturbances than the triggering condition (8.8). Besides, Table 8.1 shows that the triggering counts for the triggering condition (8.40) is smaller than that of the triggering condition (8.8). Tables 8.1–8.2 show that compared with the triggering condition (8.8) without an exponentially convergent term $\varpi_i(t)$, the two kinds of triggering conditions proposed in this chapter can reduce the amount of communication effectively and Zeno behavior does not exist.

Moreover, to illustrate the effect of the triggering parameter m_i on the consensus error, m_i is changed from $m_i = 0.01$ to $m_i = 0.1$. From Figure 8.10, it can be seen that the consensus error can be reduced by decreasing m_i.

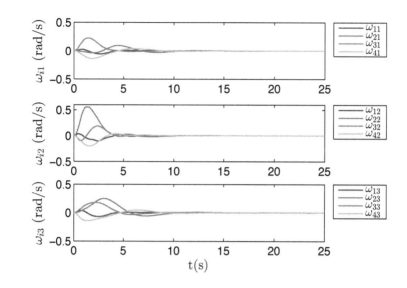

FIGURE 8.6   The states ω_i, $i = 1, \ldots, 4$.

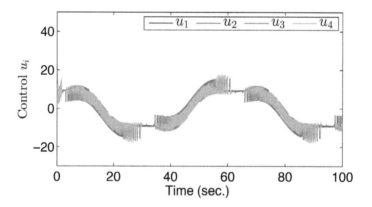

FIGURE 8.7   The control inputs u_i, $i = 1, \ldots, 4$.

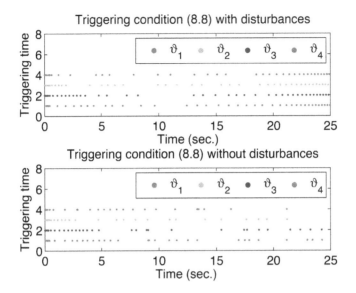

FIGURE 8.8 Comparison on the triggering times under condition (8) for the cases with and without external disturbances.

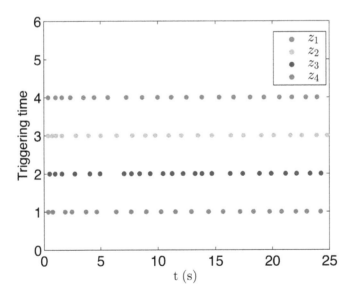

FIGURE 8.9 The triggering time instants for each agent.

TABLE 8.1

Triggering Counts of Different Triggering Conditions Proposed in this Chapter

	Condition (8.8) without $\varpi_i(t)$	Condition (8.8)	Condition (8.43)
Pendulum 1	358	37	23
Pendulum 2	462+	44	31
Pendulum 3	3479+	45	36
Pendulum 4	365+	46	14

TABLE 8.2

Inter-Event times of Different Triggering Conditions Proposed in this Chapter

		Condition (8.8) without $\varpi_i(t)$	Condition (8.8)	Condition (8.43)
Pendulum 1	Min	0.006	0.0264	0.0101
	Max	1.6595	2.6680	4.9973
Pendulum 2	Min	0.0001	0.0163	0.0096
	Max	1.6360	3.0150	3.1514
Pendulum 3	Min	0.0001	0.0163	0.0110
	Max	1.4572	4.8620	2.5098
Pendulum 4	Min	0.0001	0.0095	0.0124
	Max	0.7489	2.7920	21.1782

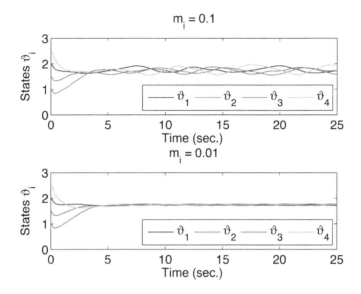

FIGURE 8.10 The states ϑ_i for $m_i = 0.1$ and $m_i = 0.01$.

8.6 NOTES

In this chapter, a fully distributed event-triggered adaptive leaderless consensus control strategy for high-order nonlinear multi-agent systems with system uncertainties is developed under a balanced digraph condition. The Lipschitz condition for nonlinear dynamics is not needed. Only one filtered signal needs to be broadcast to its neighbors for each agent. The triggering conditions are designed by using local filtered output and the latest broadcast neighboring filtered outputs such that continuous monitoring of neighboring states is not needed. Besides, to avoid the Zeno behavior, two kinds of triggering conditions are proposed.

9 Distributed Event-Triggered Adaptive Output Feedback Control

In Chapters 7 and 8, we have investigated the event-triggered consensus control problem for uncertain high-order nonlinear multi-agent systems and two consensus control schemes are presented for the leader-following and leaderless cases. However, it should be mentioned that most of currently available results are established based on the full state feedback. The results on the distributed event-triggered output feedback consensus control are still limited.

In this chapter, we investigate the distributed adaptive consensus tracking control problem for heterogenous linear multi-agent systems with unknown mismatched system parameters, event-triggered communication and directed graph via output feedback control. Due to the existence of unknown mismatched system parameters, adaptive backstepping technique is adopted as a design tool. As we know, the neighboring states are usually involved to design virtual control inputs in each step based on continuous communication, by using backstepping technique. However, for event-triggered communication case, only discrete-time neighbors' states are available in the design of virtual control inputs, which makes the virtual controllers be non-differentiable. Hence, the standard backstepping design procedure cannot be applied. To overcome this difficulty, an auxiliary system with the triggered local consensus error being its input is introduced for the agents with relative degree greater than one. Then the introduction of triggered neighbors' states can be postponed until the last step. On the other hand, to reduce the communication burden among the agents, a decentralized triggering condition is designed to transmit signals among neighbors. Unlike the triggering conditions in [137, 146] with fixed constant triggering thresholds, the triggering threshold in this chapter is time-varying and closely related to the changing rate of the signal to be transmitted. Such a triggering mechanism contributes to improving the accuracy of transmitted signal when its changing rate is small.

Then, based the designed auxiliary system and decentralized triggering condition, a novel output feedback-based distributed adaptive control scheme is proposed in this chapter. In most of the existing output feedback based consensus results, multiple signals need be transmitted among the connected agents to implement the designed consensus algorithms, such as multi-dimensional states of the dynamic compensators in [44] and [99], multi-dimensional states of the observers in [160], the local control inputs and system outputs in [136]. Different from these results, each agent needs to only transmit its one-dimensional output to its out-neighbors by using the consensus tracking control scheme presented in this chapter. This is beneficial to reduce the

DOI: 10.1201/9781003394372-9

communication burden.

9.1 PROBLEM FORMULATION

Similar to [24, 136], a group of N heterogeneous linear multi-agent systems of the following form is considered in this chapter.

$$\dot{x}_i = A_i x_i - y_i \phi_i + b_i u_i \tag{9.1}$$

$$y_i = C_i^T x_i, \qquad \text{for } i = 1, \ldots, N, \tag{9.2}$$

where $A_i = \begin{bmatrix} 0_{(n_i-1) \times 1} & I_{n_i-1} \\ 0 & 0_{1 \times (n_i-1)} \end{bmatrix}$, $b_i = \begin{bmatrix} 0_{(\rho_i-1) \times 1} \\ \bar{b}_i \end{bmatrix}$, and $C_i = [1, 0, \ldots, 0]^T \in \Re^{n_i}$. $x_i = [x_{i,1}, \ldots, x_{i,n_i}]^T \in \Re^{n_i}$, $y_i \in \Re$ and $u_i \in \Re$ are the state vector, output and control input of agent i, respectively. $0_{p \times q}$ and I_p are $p \times q$ zero matrix and $p \times p$ identity matrix, respectively. $\phi_i = [\phi_{i,n_i-1}, \ldots, \phi_{i,1}, \phi_{i,0}]^T \in \Re^{n_i}$ and $\bar{b}_i = [b_{i,m_i}, \ldots, b_{i,1}, b_{i,0}]^T \in \Re^{m_i+1}$ are vectors of unknown system parameters. ρ_i is the relative degree of agent i.

Remark 9.1 *Different from most of the existing related literature including [44, 99, 160], the system parameter vectors ϕ_i and b_i in this chapter can be totally unknown and nonidentical. Besides, the system orders and relative degree of all the agents are allowed to be distinct. Therefore, the considered linear multi-agent systems are heterogenous in both structures and parameters.*

The desired trajectory is generated by a time-varying function $y_r(t)$. Suppose that only a portion of the agents have direct access to the desired trajectory $y_r(t)$. We use $\mu_i = 1$ to denote the case that agent i can access the desired trajectory $y_r(t)$ directly, otherwise $\mu_i = 0$.

The *control objective* of this chapter is to design distributed output feedback based adaptive controllers for a group of heterogenous linear multi-agent systems as modeled in (1)–(2) with unknown system parameters and event triggering conditions for information broadcasting among neighbors via directed graph such that

- All the closed-loop signals are uniformly bounded;
- All the agents' outputs (i.e., y_i, $i \in \mathcal{V}$) can track a common desired trajectory $y_r(t)$ with a bounded tracking error $\delta_i(t) = y_i(t) - y_r(t)$, though $\mu_i = 1$ for only part of the agents.

To achieve the control objective, the following basic assumptions are imposed.

Assumption 9.1.1 *The directed graph $\mathcal{G}$ contains a spanning tree and the agent at the root node has direct access to $y_r(t)$.*

Assumption 9.1.2 *The first ρth-order derivatives of the desired trajectory y_r are bounded and piecewise continuous, where $\rho = \max\{\rho_1, \ldots, \rho_N\}$.*

Assumption 9.1.3 *The polynomial $N_i(s) = b_{i,m_i} s^{m_i} + \ldots + b_{i,1} s + b_{i,0}$ is Hurwitz. Besides, the system order n_i, m_i, and the sign of b_{i,m_i} are known by each agent i.*

9.2 DESIGN OF DISTRIBUTED ADAPTIVE CONTROLLERS WITH EVENT-TRIGGERED COMMUNICATION

9.2.1 DESIGN OF LOCAL STATE ESTIMATION FILTERS

For each agent i, the following local filters are designed to estimate its unmeasurable states [56, Chapter 9]

$$\dot{\eta}_i = A_{i0}\eta_i + e_{n_i,n_i}y_i, \tag{9.3}$$

$$\dot{\lambda}_i = A_{i0}\lambda_i + e_{n_i,n_i}u_i, \tag{9.4}$$

where $\eta_i = [\eta_{i,1}, \ldots, \eta_{i,n_i}]^T \in \Re^{n_i}$ and $\lambda_i = [\lambda_{i,1}, \ldots, \lambda_{i,n_i}]^T \in \Re^{n_i}$ are the state vectors of local state estimation filters. In this chapter, $e_{p,q}$ is used to denote the qth coordinate vector in $\Re^p$. A_{i0} is a Hurwitz matrix of the form $A_{i0} = A_i - k_i e_{n_i,1}{}^T$, where $k_i = [k_{i,1}, k_{i,2}, \ldots, k_{i,n_i}]^T$ is a constant vector chosen to render A_{i0} Hurwitz. Since A_{i0} is Hurwitz, there exists a matrix positive definite matrix $\Phi_i \in \Re^{n_i \times n_i}$ such that $\Phi_i A_{i0} + A_{i0}{}^T \Phi_i = -I_{n_i}$.

Define a set of column vectors as $\xi_i = [\xi_{i,1}, \ldots, \xi_{i,n_i}]^T = -A_{i0}^{n_i}\eta_i$, $v_{il} = [v_{il,1}, \ldots, v_{il,n_i}]^T = A_{i0}^l \lambda_i$, $l = 0, \ldots, m_i$ and two matrices as $\Xi_i = -[A_{i0}^{n_i-1}\eta_i, \ldots, A_{i0}\eta_i, \eta_i]$, $\Omega_i^T = [v_{im_i}, \ldots, v_{i1}, v_{i0}, \Xi_i]$. Let $\hat{x}_i$ be the estimate of unmeasurable state x_i with an estimation error as $\epsilon_i = x_i - \hat{x}_i$. Define $\hat{x}_i = \xi_i + \Omega_i^T\theta_i$ with $\theta_i = [b_i^T, \phi_i^T]^T$. Then, the static relationship between the state x_i and the unknown parameter θ_i can be expressed as $x_i = \xi_i + \Omega_i^T\theta_i + \epsilon_i$. Combining (9.3),(9.4), ξ_i, v_{il}, Ξ_i, and Ω_i, it can be shown that ϵ_i is governed by $\dot{\epsilon}_i = A_{i0}\epsilon_i$, which shows that the state estimation error ϵ_i will exponentially decay to zero as time progresses [56].

Let $\omega_i = [v_{im_i,2}, v_{i(m_i-1),2}, \ldots, v_{i0,2}, \Xi_{i(2)} - y_i e_{n_i,1}^T]^T$ and $\bar{\omega}_i = [0, v_{i(m_i-1),2}, \ldots, v_{i0,2}, \Xi_{i(2)} - y_i e_{n_i,1}^T]^T$. Since $x_i = \xi_i + \Omega_i^T\theta_i + \epsilon_i$, we have $x_{i,2} = \xi_{i,2} + \omega_i^T\theta_i + \epsilon_{i,2}$, where $\epsilon_{i,2}$ is the second entry of vector ϵ_i. From (9.1) and (9.2), we can further get $\dot{y}_i = b_{i,m_i}v_{im_i,2} + \xi_{i,2} + \bar{\omega}_i^T\theta_i + \epsilon_{i,2}$. With the designed filters (9.3),(9.4), and following the similar development in [56], the system (9.1) can be transformed into

$$\dot{y}_i = b_{i,m_i}v_{im_i,2} + \xi_{i,2} + \bar{\omega}_i^T\theta_i + \epsilon_{i,2}$$
$$\dot{v}_{im_i,2} = v_{im_i,3} - k_{i,2}v_{im_i,1},$$
$$\vdots$$
$$\dot{v}_{im_i,\rho_i-1} = v_{im_i,\rho_i} - k_{i,\rho_i-1}v_{im_i,1},$$
$$\dot{v}_{im_i,\rho_i} = v_{im_i,\rho_i+1} - k_{i,\rho_i}v_{im_i,1} + u_i. \tag{9.5}$$

Note that all the states of system (9.5) are available for feedback control design.

9.2.2 DESIGN OF TIME-VARYING TRIGGERING RULES

For agent i, the sequence of event times is denoted by $t_0^i, t_1^i, \ldots, t_{k_i}^i, \ldots, k_i \in Z^+$, at which agent i broadcasts its output $y_i(t_{k_i}^i)$ to its out-neighbors. t_0^i is the initial time instant when agent i starts up. We denote the latest broadcast output by $\bar{y}_i(t) = y_i(t^i_{k_i})$ for $t \in [t^i_{k_i}, t^i_{k_i+1})$, $i \in \mathcal{V}$. In each agent, a novel decentralized triggering condition with time-varying triggering threshold is designed as

$$t^i_{k_i+1} = \inf \left\{ t > t^i_{k_i} \mid |y_i(t) - y_i(t^i_{k_i})| \geq \Pi_i(t) \right\} \tag{9.6}$$

with

$$\Pi_i(t) = \begin{cases} \frac{\pi_{i,2}-\pi_{i,1}}{M_i} R_i(t) + \pi_{i,1}, & \text{if } R_i(t) \leq M_i \\ \pi_{i,2}, & \text{otherwise} \end{cases}, \tag{9.7}$$

where $R_i(t) = \frac{|y_i(t)-y_i(t^i_{k_i})|}{t-t^i_{k_i}}$ for $t > 0$ and $R_i(0) = 0$, which stands for the absolute value of the changing rate of $y_i(t)$ approximately. $\pi_{i,1}$, $\pi_{i,2}$, and M_i are positive design parameters and $\pi_{i,2} > \pi_{i,1}$. M_i determines the adjustable range of the triggering threshold. $\pi_{i,1}$ is added to avoid the Zeno behavior.

Remark 9.2 *We now recall the triggering condition with a positive constant triggering threshold as expressed by $t^i_{k_i+1} = \inf \left\{ t > t^i_{k_i} \mid |y_i(t) - y_i(t^i_{k_i})| \geq \pi_{i,2} \right\}$. Such a triggering condition is commonly adopted in the literature including [137, 146]. It is beneficial to avoid the occurrence of Zeno behavior, while continuous monitoring of neighbors' states is not needed. In implementation, a large fixed constant triggering threshold $\pi_{i,2}$ needs often be chosen to reduce the communication burden. However, a large $\pi_{i,2}$ normally leads to signal distortion especially around the extremum points of the transmitting signal when the signal's changing rate is small. To handle this issue, an improved version is proposed in this chapter. It can be observed from (9.7) that the triggering threshold $\Pi_i(t)$ is a time-varying function $\frac{\pi_{i,2}-\pi_{i,1}}{M_i} R_i(t) + \pi_{i,1}$ for $R_i(t) \leq M_i$, which varies according to the size of $R_i(t)$. The smaller $R_i(t)$ results in a smaller triggering threshold $\Pi_i(t)$. As a result, the triggering times of transmitting signal $y_i(t)$ increase, then the aforementioned distortion problem can be ameliorated.*

9.2.3 DESIGN OF DISTRIBUTED ADAPTIVE CONTROLLERS

To design distributed adaptive consensus controllers, the following error variables are firstly introduced for each agent.

$$z_{i,1} = y_i - \mu_i y_r - (1 - \mu_i)\hat{y}_{i,r} \tag{9.8}$$

$$= \delta_i + (1 - \mu_i)\tilde{y}_{i,r}, \tag{9.9}$$

$$\zeta_i = \sum_{j=1}^{N} a_{ij}(y_i - y_j) + \mu_i(y_i - y_r), \tag{9.10}$$

$$\bar{\zeta}_i = \sum_{j=1}^{N} a_{ij}(y_i - \bar{y}_j) + \mu_i(y_i - y_r), \tag{9.11}$$

where $\tilde{y}_{i,r} = y_r - \hat{y}_{i,r}$ and $\hat{y}_{i,r}$ is an estimate of y_r, which is introduced only for agent i with $\mu_i = 0$. ζ_i and $\bar{\zeta}_i$ represent the actual local consensus error and the consensus error that is computable with event-triggered communication, respectively. $\delta_i = y_i - y_r$ denotes the actual tracking error.

9.2.3.1 Agents with the Relative Degree $\rho_i = 1$

In this part, we consider the agents with the relative degree of one, i.e., $\rho_i = 1, i \in \mathcal{V}$. From (9.5) and (9.8), the derivative of $z_{i,1}$ is computed as

$$\dot{z}_{i,1} = b_{i,m_i} u_i + \xi_{i,2} + \omega_i^T \theta_i + \epsilon_{i,2} - \mu_i \dot{y}_r - (1 - \mu_i)\dot{\hat{y}}_{i,r}. \tag{9.12}$$

The distributed adaptive control input u_i is designed as

$$u_i = \hat{\varrho}_i \bar{\alpha}_{i,1}, \tag{9.13}$$

$$\bar{\alpha}_{i,1} = - k\hat{P}_i \bar{\zeta}_i - c_{i,1} z_{i,1} - d_{i,1} z_{i,1} - \xi_{i,2} - \omega_i^T \hat{\theta}_i$$
$$+ \mu_i \dot{y}_r - \tanh\left(\frac{z_{i,1}}{\varepsilon_i}\right)\hat{D}_i + (1 - \mu_i)\dot{\hat{y}}_{i,r}, \tag{9.14}$$

where k, $c_{i,1}$, $d_{i,1}$, and ε_i are positive design parameters. $\hat{\varrho}_i$, $\hat{P}_i$, $\hat{\theta}_i$, and $\hat{D}_i$ are the estimates of unknown parameters $\varrho_i = \frac{1}{b_{i,m_i}}$, P_i defined in Lemma 2.4, θ_i and $D_i = kP_i\Delta_{\zeta_i}$ with Δ_{ζ_i} being a positive constant, respectively. The detailed definition of Δ_{ζ_i} will be given in later discussion.

Based on Lyapunov design approach, the parameter update laws for $\hat{\varrho}_i$, $\hat{P}_i$, $\hat{D}_i$, $\hat{y}_{i,r}$, and $\hat{\theta}_i$ are designed as

$$\dot{\hat{\varrho}}_i = -\gamma_{\varrho_i} \text{sgn}(b_{i,m_i})\bar{\alpha}_{i,1} z_{i,1} - \gamma_{\varrho_i}\kappa_{\varrho_i}(\hat{\varrho}_i - \varrho_{i,0}), \tag{9.15}$$

$$\dot{\hat{P}}_i = \gamma_{P_i} \bar{\zeta}_i z_{i,1} - \gamma_{P_i}\kappa_{P_i}(\hat{P}_i - P_{i,0}), \tag{9.16}$$

$$\dot{\hat{D}}_i = \gamma_{D_i} z_{i,1} \tanh\left(\frac{z_{i,1}}{\varepsilon_i}\right) - \gamma_{D_i}\kappa_{D_i}(\hat{D}_i - D_{i,0}), \tag{9.17}$$

$$\dot{\hat{y}}_{i,r} = -\gamma_{y_i} \bar{\zeta}_i - \gamma_{y_i}\kappa_{y_i}(\hat{y}_{i,r} - y_{i,0}), \tag{9.18}$$

$$\dot{\hat{\theta}}_i = \Gamma_i \omega_i z_{i,1} - \Gamma_i \kappa_{\theta_i}(\hat{\theta}_i - \theta_{i,0}), \tag{9.19}$$

where $\text{sgn}(\cdot)$ and $\tanh(\cdot)$ denote the sign function and hyperbolic tangent function, respectively. γ_{ϱ_i}, γ_{P_i}, γ_{D_i}, γ_{y_i}, Γ_i, κ_{ϱ_i}, κ_{P_i}, κ_{D_i}, κ_{y_i}, κ_{θ_i}, $\varrho_{i,0}$, $P_{i,0}$, $D_{i,0}$, $y_{i,0}$, and $\theta_{i,0}$ are all positive design parameters with appropriate dimension.

Let the estimation errors for the above unknown parameters be $\tilde{\varrho}_i = \varrho_i - \hat{\varrho}_i$, $\tilde{P}_i = P_i - \hat{P}_i$, $\tilde{D}_i = D_i - \hat{D}_i$ and $\tilde{\theta}_i = \theta_i - \hat{\theta}_i$. Define a Lyapunov function $V_{i,1}$ as

$$V_{i,1} = \frac{1}{2}z_{i,1}^2 + \frac{1}{2}\tilde{\theta}_i^T \Gamma_i^{-1}\tilde{\theta}_i + \frac{|b_{i,m_i}|}{2\gamma_{\varrho_i}}\tilde{\varrho}_i^2 + \frac{k}{2\gamma_{P_i}}\tilde{P}_i^2$$
$$+ \frac{1}{2\gamma_{D_i}}\tilde{D}_i^2 + \frac{kP_i(1 - \mu_i)}{2\gamma_{y_i}}\tilde{y}_{i,r}^2 + \frac{\epsilon_i^T \Phi_i \epsilon_i}{2d_{i,1}}. \tag{9.20}$$

From (9.9), (9.12)–(9.14), the derivative of $V_{i,1}$ can be derived as

$$\dot{V}_{i,1} \leq -kP_i\bar{\zeta}_i\delta_i - z_{i,1}\tanh\left(\frac{z_{i,1}}{\varepsilon_i}\right)D_i - c_{i,1}z_{i,1}^2 + \frac{k}{\gamma_{P_i}}\tilde{P}_i\left(\gamma_{P_i}\bar{\zeta}_i z_{i,1} - \dot{P}_i\right)$$

$$+ \frac{|b_{i,m_i}|}{\gamma_{\varrho_i}}\tilde{\varrho}_i\left[-\text{sgn}(b_{i,m_i})\gamma_{\varrho_i}\bar{\alpha}_{i,1}z_{i,1} - \dot{\hat{\varrho}}_i\right] + \tilde{\theta}_i^T\Gamma_i^{-1}\left(\Gamma_i\omega_i z_{i,1} - \dot{\hat{\theta}}_i\right)$$

$$+ \frac{1}{\gamma_{D_i}}\tilde{D}_i\left[\gamma_{D_i}z_{i,1}\tanh\left(\frac{z_{i,1}}{\varepsilon_i}\right) - \dot{\hat{D}}_i\right] - \frac{1}{4d_{i,1}}\epsilon_i^T\epsilon_i$$

$$+ \frac{kP_i(1-\mu_i)}{\gamma_{y_i}}\tilde{y}_{i,r}\left(\dot{y}_r - \gamma_{y_i}\bar{\zeta}_i z_{i,1} - \dot{\hat{y}}_{i,r}\right) \tag{9.21}$$

With the triggering mechanism (9.6), it can be obtained that $|y_i(t) - \bar{y}_i(t)| \leq \Pi_i(t)$. From (9.10) and (9.11), we have $\bar{\zeta}_i(t) = \zeta_i(t) + \sum_{j=1}^{N} a_{ij}\Pi_j(t)\iota_{\zeta_j}(t)$, where $\iota_{\zeta_j}(t)$ is a time-varying function with $|\iota_{\zeta_j}(t)| \leq 1$. Denote the upper bounds of $\dot{y}_r$ and $\sum_{j=1}^{N} a_{ij}\Pi_j(j)\iota_{\zeta_j}(t)$ by $\bar{\dot{y}}_r$ and Δ_{ζ_i}, respectively. Substituting (9.15)–(9.19) into (9.21), we can further obtain

$$\dot{V}_{i,1} \leq -kP_i\bar{\zeta}_i\delta_i - c_{i,1}z_{i,1}^2 - \frac{|b_{i,m_i}|\kappa_{\varrho_i}}{2}\tilde{\varrho}_i^2 - \frac{\kappa_{D_i}}{2}\tilde{D}_i^2 - \frac{k\kappa_{P_i}}{2}\tilde{P}_i^2$$

$$- \frac{kP_i(1-\mu_i)\kappa_{y_i}}{4}\tilde{y}_{i,r}^2 - \frac{\kappa_{\theta_i}\|\tilde{\theta}_i\|^2}{2} - \frac{1}{4d_{i,1}}\epsilon_i^T\epsilon_i + M_{i,1}, \tag{9.22}$$

where $M_{i,1} = \frac{|b_{i,m_i}|\kappa_{\varrho_i}}{2}(\varrho_i - \varrho_{i,0})^2 + \frac{\kappa_{D_i}}{2}(D_i - D_{i,0})^2 + \frac{k\kappa_{P_i}}{2}(P_i - P_{i,0})^2 + 0.2785\varepsilon_i D_i + \frac{kP_i(1-\mu_i)\kappa_{y_i}}{2}(y_{i,r} - y_{i,0})^2 + \frac{kP_i(1-\mu_i)}{\kappa_{y_i}\gamma_{y_i}^2}(\bar{\dot{y}}_r + \gamma_{y_i}\Delta_{\zeta_i})^2 + \frac{\kappa_{\theta_i}\|\theta_i - \theta_{i,0}\|^2}{2}$. Note that the Young's inequality $|ab| \leq a^2 + \frac{b^2}{4}$ for $a, b \in \Re$ and $0 \leq |z_{i,1}| - z_{i,1}\tanh\left(\frac{z_{i,1}}{\varepsilon_i}\right) \leq 0.2785\varepsilon_i$ have been used.

9.2.3.2 Agents with the Relative Degree $\rho_i \geq 2$

In this part, we focus on the agents whose relative degree is greater than one, i.e., $\rho_i \geq 2$, $i \in \mathcal{V}$. To account for unmatched unknown parameters, adaptive backstepping technique [56] is employed. Except for the error variables defined in (9.8)–(9.11), the following change of coordinates is augmented.

$$z_{i,q} = v_{im_i,q} - \alpha_{i,q-1}, q = 2, \ldots, \rho_i, \tag{9.23}$$

where $\alpha_{i,q-1}, q = 2, \ldots, \rho_i$, are the virtual control inputs to be designed recursively. We now proceed with the design of distributed adaptive controllers step by step.

Step 0. For agent i with $\rho_i \geq 2$, we first introduce the following $(\rho_i - 1)$th-order auxiliary system with the triggered consensus error $\bar{\zeta}_i$ being its input.

$$\dot{s}_{i,q} = -\tau_{i,q}s_{i,q} + \tau_{i,q}s_{i,q+1}, q = 1, \ldots, \rho_i - 2,$$

$$\dot{s}_{i,\rho_i-1} = -\tau_{i,\rho_i-1}s_{i,\rho_i-1} + \tau_{i,\rho_i-1}\bar{\zeta}_i, \tag{9.24}$$

where $s_{i,1}, \ldots, s_{i,\rho_i-1}$ are the states of auxiliary system (9.24), $\tau_{i,\rho_i-1} = 3 + \tau_{i,0}$, $\tau_{i,\rho_i-2} = \frac{1}{4}\tau_{i,\rho_i-1}^2 + 3 + \tau_{i,0}$, $\tau_{i,\rho_i-3} = \frac{1}{4}\tau_{i,\rho_i-2}^2 + 2 + \tau_{i,0}, \ldots, \tau_{i,2} = \frac{1}{4}\tau_{i,3}^2 + 2 + \tau_{i,0}$, and $\tau_{i,1} = \frac{1}{4}\tau_{i,2}^2 + 1 + \tau_{i,0}$ are positive design parameters, where $\tau_{i,0}$ is a positive constant. The initial states of auxiliary system (9.24) are set as $s_{i,1}(0) = \ldots = s_{i,\rho_i-1}(0) = \zeta_i(0)$ to reduce the initial error between $s_{i,1}$ and ζ_i. With $\bar{\zeta}_i(t) = \zeta_i(t) + \sum_{j=1}^{N} a_{ij}\Pi_j(t)\iota_{\zeta_j}(t)$, the last equation of (9.24) can be rewritten as $\dot{s}_{i,\rho_i-1} = -\tau_{i,\rho_i-1}s_{i,\rho_i-1} + \tau_{i,\rho_i-1}[\zeta_i(t) + d_i(t)]$, where $d_i(t) = \sum_{j=1}^{N} a_{ij}\Pi_j(t)\iota_{\zeta_j}(t)$. Since $\Pi_j(t)$ and $\iota_{\zeta_j}(t)$ are bounded time-varying signals, $d_i(t)$ can be considered as a bounded disturbance with an upper bound $\bar{d}_i = \sum_{j=1}^{N} a_{ij}\pi_{j,2}$. We define a set of tracking error variables as $\tilde{s}_{i,\rho_i-1} = s_{i,\rho_i-1} - \zeta_i$, $\tilde{s}_{i,\rho_i-2} = s_{i,\rho_i-2} - s_{i,\rho_i-1}, \ldots, \tilde{s}_{i,1} = s_{i,1} - s_{i,2}$, and $V_{i,s} = \frac{1}{2}\sum_{l=1}^{\rho_i-1}\tilde{s}_{i,l}^2$. From (9.24) and using Young's inequality, the derivative of $V_{i,s}$ can be computed as

$$\begin{aligned}
\dot{V}_{i,s} \leq & -(\tau_{i,\rho_i-1} - 2)\tilde{s}_{i,\rho_i-1}^2 + \frac{1}{4}(\tau_{i,\rho_i-1}d_i - \dot{\zeta}_i)^2 \\
& -(\tau_{i,\rho_i-2} - \frac{1}{4}\tau_{i,\rho_i-1}^2 - 2)\tilde{s}_{i,\rho_i-2}^2 + \frac{1}{4}(\tau_{i,\rho_i-1}d_i)^2 \\
& -\sum_{l=2}^{\rho_i-3}(\tau_{i,l} - \frac{1}{4}\tau_{i,l+1}^2 - 1)\tilde{s}_{i,l}^2 - (\tau_{i,1} - \frac{1}{4}\tau_{i,2}^2)\tilde{s}_{i,1}^2 \\
\leq & -\sum_{l=1}^{\rho_i-1}(\tau_{i,0} + 1)\tilde{s}_{i,l}^2 + M_{i,s}, \tag{9.25}
\end{aligned}$$

where $M_{i,s} = \frac{1}{4}(\tau_{i,\rho_i-1}\bar{d}_i + |\dot{\zeta}_i|)^2 + \frac{1}{4}(\tau_{i,\rho_i-1}\bar{d}_i)^2$.

It can be observed from (9.25) that $\dot{V}_{i,s} \leq 0$ when $\sum_{l=1}^{\rho_i-1}\tilde{s}_{i,l}^2 \geq \frac{M_{i,s}}{\tau_{i,0}+1}$. Note that $\bar{d}_i$ can be reduced by decreasing $\pi_{i,2}$, thereby $M_{i,s}$ can be reduced. Since

$$|s_{i,1} - \zeta_i| \leq \sqrt{\left(\sum_{l=1}^{\rho_i-1}|\tilde{s}_{i,l}|\right)^2} \leq \sqrt{(\rho_i-1)\sum_{l=1}^{\rho_i-1}\tilde{s}_{i,l}^2},$$

the tracking error $|s_{i,1} - \zeta_i|$ can be reduced by decreasing $\pi_{i,2}$.

Step 1. From (9.5), (9.8) and (9.23) for $q = 2$, the derivative of $z_{i,1}$ is

$$\begin{aligned}
\dot{z}_{i,1} = & b_{i,m_i}z_{i,2} + b_{i,m_i}\alpha_{i,1} + \xi_{i,2} + \bar{\omega}_i^T\theta_i + \epsilon_{i,2} - \mu_i\dot{y}_r \\
& -(1 - \mu_i)\dot{\hat{y}}_{i,r}. \tag{9.26}
\end{aligned}$$

The first virtual control input $\alpha_{i,1}$ is chosen as

$$\alpha_{i,1} = \hat{\varrho}_i\bar{\alpha}_{i,1}, \tag{9.27}$$

$$\begin{aligned}
\bar{\alpha}_{i,1} = & -k\hat{P}_is_{i,1} - c_{i,1}z_{i,1} - d_{i,1}z_{i,1} - \xi_{i,2} - \bar{\omega}_i^T\hat{\theta}_i \\
& + \mu_i\dot{y}_r + (1 - \mu_i)\dot{\hat{y}}_{i,r}, \tag{9.28}
\end{aligned}$$

where the design parameters k, $c_{i,1}$, $d_{i,1}$ and the estimator $\hat{\varrho}_i$, $\hat{P}_i$, $\hat{\theta}_i$ are defined in the same way as those in Section 9.2.3.1.

The parameter update laws for $\hat{\varrho}_i$, $\hat{P}_i$, and $\hat{y}_{i,r}$ are chosen as

$$\dot{\hat{\varrho}}_i = -\gamma_{\varrho_i}\mathrm{sgn}(b_{i,m_i})\bar{\alpha}_{i,1}z_{i,1} - \gamma_{\varrho_i}\kappa_{\varrho_i}(\hat{\varrho}_i - \varrho_{i,0}), \tag{9.29}$$

$$\dot{\hat{P}}_i = \gamma_{P_i}s_{i,1}z_{i,1} - \gamma_{P_i}\kappa_{P_i}(\hat{P}_i - P_{i,0}), \tag{9.30}$$

$$\dot{\hat{y}}_{i,r} = -\gamma_{y_i}s_{i,1} - \gamma_{y_i}\kappa_{y_i}(\hat{y}_{i,r} - y_{i,0}), \tag{9.31}$$

where the design parameters are defined in the same way as those in (9.15)–(9.18). Let $\chi_{i,1} = \omega_i - \alpha_{i,1}e_{n_i+m_i+1,1}$. Substituting (9.27) and (9.28) into (9.26) gets

$$\dot{z}_{i,1} = \hat{b}_{i,m_i}z_{i,2} + \chi_{i,1}^T\tilde{\theta}_i - kP_is_{i,1} + k\tilde{P}_is_{i,1} - c_{i,1}z_{i,1}$$
$$- d_{i,1}z_{i,1} - b_{i,m_i}\tilde{\varrho}_i\bar{\alpha}_{i,1} + \epsilon_{i,2}, \tag{9.32}$$

where $\hat{b}_{i,m_i} = \hat{\theta}_{i,1}$ is the estimate of b_{i,m_i} and $\hat{\theta}_{i,1}$ is the first element of $\hat{\theta}_i$. Define $\Upsilon_{i,1} = \chi_{i,1}z_{i,1}$. Choose a Lyapunov function $V_{i,1}$ as

$$V_{i,1} = \frac{1}{2}z_{i,1}^2 + \frac{1}{2}\tilde{\theta}_i^T\Gamma_i^{-1}\tilde{\theta}_i + \frac{|b_{i,m_i}|}{2\gamma_{\varrho_i}}\tilde{\varrho}_i^2 + \frac{k}{2\gamma_{P_i}}\tilde{P}_i^2$$
$$+ \frac{kP_i(1-\mu_i)}{2\gamma_{y_i}}\tilde{y}_{i,r}^2 + \frac{\epsilon_i^T\Phi_i\epsilon_i}{2d_{i,1}} + V_{i,s}. \tag{9.33}$$

With $s_{i,1} = \zeta_i + \tilde{s}_{i,1} + \ldots + \tilde{s}_{i,\rho_i-1}$ and (9.27)–(9.32), the derivative of $V_{i,1}$ can be derived as

$$\dot{V}_{i,1} \leq \hat{b}_{i,m_i}z_{i,2}z_{i,1} - kP_i[\zeta_i + \tilde{s}_{i,1} + \ldots + \tilde{s}_{i,\rho_i-1}]\delta_i$$
$$+ |b_{i,m_i}|\kappa_{\varrho_i}\tilde{\varrho}_i(\hat{\varrho}_i - \varrho_{i,0}) + k\kappa_{P_i}\tilde{P}_i(\hat{P}_i - P_{i,0})$$
$$+ \frac{kP_i(1-\mu_i)}{\gamma_{y_i}}\tilde{y}_{i,r}\dot{y}_r + \tilde{\theta}_i^T\Gamma_i^{-1}\left(\Gamma_i\Upsilon_{i,1} - \dot{\hat{\theta}}_i\right)$$
$$+ kP_i(1-\mu_i)\kappa_{y_i}\tilde{y}_{i,r}(\hat{y}_{i,r} - y_{i,0}) - \frac{1}{4d_{i,1}}\epsilon_i^T\epsilon_i$$
$$- c_{i,1}z_{i,1}^2 + \dot{V}_{i,s}$$
$$\leq \hat{b}_{i,m_i}z_{i,2}z_{i,1} - kP_i\zeta_i\delta_i - \frac{|b_{i,m_i}|\kappa_{\varrho_i}}{2}\tilde{\varrho}_i^2 - \frac{k\kappa_{P_i}}{2}\tilde{P}_i^2$$
$$- \frac{kP_i(1-\mu_i)\kappa_{y_i}}{4}\tilde{y}_{i,r}^2 - \frac{1}{4d_{i,1}}\epsilon_i^T\epsilon_i - c_{i,1}z_{i,1}^2 + M_{i,2}$$
$$+ \tilde{\theta}_i^T\Gamma_i^{-1}\left(\Gamma_i\Upsilon_{i,1} - \dot{\hat{\theta}}_i\right) - \sum_{l=1}^{\rho_i-1}\tau_{i,0}\tilde{s}_{i,l}^2, \tag{9.34}$$

where $M_{i,2} = \frac{|b_{i,m_i}|\kappa_{\varrho_i}}{2}(\varrho_i - \varrho_{i,0})^2 + \frac{k\kappa_{P_i}}{2}(P_i - P_{i,0})^2 + \frac{kP_i(1-\mu_i)\kappa_{y_i}}{2}(y_{i,r} - y_{i,0})^2 + \frac{kP_i(1-\mu_i)}{\kappa_{y_i}\gamma_{y_i}^2}\tilde{y}_r^2 + \frac{\rho_i-1}{4}(kP_i\delta_i)^2 + M_{i,s}$.

Remark 9.3 *For the agent with the relative degree $\rho_i \geq 2$, if the virtual control input $\bar{\alpha}_{i,1}$ is designed as that in (9.14), $\bar{\zeta}_i = \sum_{j=1}^{N} a_{ij}(y_i - \bar{y}_j) + \mu_i(y_i - y_r)$ containing the triggered neighboring outputs $\bar{y}_j, j \in \mathcal{N}_i$ is involved. Since $\bar{y}_j$ is a piecewise continuous signal, $\bar{\alpha}_{i,1}$ is non-differentiable. It will make the backstepping design procedure impossible to proceed. To address this issue, we introduce a $(\rho_i - 1)$th-order auxiliary system (9.24) firstly. Since the state $s_{i,1}$ is $(\rho_i - 1)$th-order differentiable, we then replace the signal $\bar{\zeta}_i$ with $s_{i,1}$ in the design of $\bar{\alpha}_{i,1}$ in (9.28).*

Step q ($q = 2, ..., \rho_i - 1$). We take $\alpha_{i,1}$ as an example to clarify the arguments of virtual control inputs. From the definition of ξ_i and Ξ_i, $\xi_{i,2}$ and $\Xi_{i(2)}$ are the function of η_i. Let $\bar{\lambda}_{i,l} = [\lambda_{i,1}, \ldots, \lambda_{i,l}]^T$ and $\lambda_{i,l} = 0$ for $l > \rho_i$. Similar to [56], it can be shown that $v_{il,2}$ can be expressed as $v_{il,2} = [*, \ldots, *, 1]\bar{\lambda}_{i,l+2}$, where $*$ denotes the element that has any value. Therefore, by examining (9.27) and (9.28), $\alpha_{i,1}$ is the function of $\hat{\varrho}_i$, $\hat{P}_i$, $s_{i,1}$, y_i, $\mu_i y_r$, $\mu_i \dot{y}_r$, $(1 - \mu_i)\hat{y}_{i,r}$, η_i, $\bar{\lambda}_{i,m_i+1}$, and $\hat{\theta}_i$. Then following the similar analysis, the arguments of $\alpha_{i,q}$, $q = 2, \ldots, \rho_i - 1$, can also be clarified in the sequel.

Define $\Upsilon_{i,q} = \Upsilon_{i,q-1} - \frac{\partial \alpha_{i,q-1}}{\partial y_i} \omega_i z_{i,q}$. The virtual control input $\alpha_{i,q}$ is chosen as

$$
\begin{aligned}
\alpha_{i,q} = & -z_{i,q-1} - c_{i,q} z_{i,q} - d_{i,q}\left(\frac{\partial \alpha_{i,q-1}}{\partial y}\right)^2 z_{i,q} \\
& + k_{i,q} v_{im_i,1} + \frac{\partial \alpha_{i,q-1}}{\partial \hat{\varrho}_i} \dot{\hat{\varrho}}_i + \frac{\partial \alpha_{i,q-1}}{\partial \hat{P}_i} \dot{\hat{P}}_i \\
& + \sum_{l=1}^{q-1} \frac{\partial \alpha_{i,q-1}}{\partial s_{i,l}}(-\tau_i s_{i,l} + \tau_i s_{i,l+1}) + \frac{\partial \alpha_{i,q-1}}{\partial y_i}\xi_{i,2} \\
& + \mu_i \sum_{l=1}^{q} \frac{\partial \alpha_{i,q-1}}{\partial y_r^{(l-1)}} y_r^{(l)} + (1 - \mu_i)\frac{\partial \alpha_{i,q-1}}{\partial \hat{y}_{i,r}} \dot{\hat{y}}_{i,r} \\
& + \frac{\partial \alpha_{i,q-1}}{\partial \eta_i}(A_{i,0}\eta_i + e_{n_i n_i} y_i) \\
& + \sum_{l=1}^{m_i+q-1} \frac{\partial \alpha_{i,q-1}}{\partial \lambda_{i,l}}(-k_{i,l}\lambda_{i,1} + \lambda_{i,l+1}) \\
& + \frac{\partial \alpha_{i,q-1}}{\partial y_i}\omega_i^T \hat{\theta}_i - \left(\sum_{l=2}^{q-1} z_{i,l} \frac{\partial \alpha_{i,l-1}}{\partial \hat{\theta}_i}\right)\Gamma_i \frac{\partial \alpha_{i,q-1}}{\partial y}\omega_i \\
& + \frac{\partial \alpha_{i,q-1}}{\partial \hat{\theta}_i}\Gamma_i\left(\Upsilon_{i,q} - \kappa_{i,\theta}(\hat{\theta}_i - \theta_{i,0})\right),
\end{aligned}
\tag{9.35}
$$

where $c_{i,q}$ and $d_{i,q}$ are positive design parameters. Note that for the virtual control input $\alpha_{i,2}$, the term $-z_{i,q-1}$ should be replaced by $-\hat{b}_{i,m_i} z_{i,1}$ and the term $-\left(\sum_{l=2}^{q-1} z_{i,l} \frac{\partial \alpha_{i,l-1}}{\partial \hat{\theta}_i}\right)\Gamma_i \frac{\partial \alpha_{i,q-1}}{\partial y}\omega_i$ is not needed.

The Lyapunov function $V_{i,q}$ is defined as $V_{i,q} = V_{i,q-1} + \frac{1}{2}z_{i,q}^2 + \frac{\epsilon_i^T \Phi_i \epsilon_i}{2d_{i,q}}$, whose derivative can be derived as

$$
\dot{V}_{i,q} \leq z_{i,q} z_{i,q+1} - kP_i \zeta_i \delta_i - \sum_{l=1}^{q} c_{i,l} z_{i,l}^2 - \frac{|b_{i,m_i}| \kappa_{\varrho_i}}{2} \tilde{\varrho}_i^2
$$

$$
- \sum_{l=1}^{\rho_i - 1} \tau_{i,0} \tilde{s}_{i,l}^2 - \frac{k\kappa_{P_i}}{2} \tilde{P}_i^2 - \frac{kP_i(1 - \mu_i)\kappa_{y_i}}{4} \tilde{y}_{i,r}^2
$$

$$
- \sum_{l=1}^{q} \frac{1}{4d_{i,l}} \epsilon_i^T \epsilon_i + M_{i,1} + \tilde{\theta}_i^T \Gamma_i^{-1} \left(\Gamma_i \Upsilon_{i,q} - \dot{\hat{\theta}}_i \right)
$$

$$
- \left(\sum_{l=2}^{q-1} z_{i,q} \frac{\partial \alpha_{i,q-1}}{\partial \hat{\theta}_i} \right) \left[\dot{\hat{\theta}}_i - \Gamma_i \left(\Upsilon_{i,q} - \kappa_{i,\theta}(\hat{\theta}_i - \theta_{i,0}) \right) \right]. \tag{9.36}
$$

Step ρ_i. Define $\Upsilon_{i,\rho_i} = \Upsilon_{i,\rho_i-1} - \frac{\partial \alpha_{i,\rho_i-1}}{\partial y_i} w_i z_{i,\rho_i}$. The distributed adaptive control input u_i is designed as

$$
u_i = -v_{im_i,\rho_i+1} + \alpha_{i,\rho_i}, \tag{9.37}
$$

where α_{i,ρ_i} is defined in (9.35) for $q = \rho_i$ and $s_{i,\rho_i} = \bar{\zeta}_i$. Note that for the agents with $n_i = \rho_i$, the term $-v_{im_i,\rho_i+1}$ does not exist.

The parameter update law for $\hat{\theta}_i$ is chosen as

$$
\dot{\hat{\theta}}_i = \Gamma_i \Upsilon_{i,\rho_i} - \Gamma_i \kappa_{\theta_i}(\hat{\theta}_i - \theta_{i,0}), \tag{9.38}
$$

where Γ_i, κ_{θ_i}, $\theta_{i,0}$ are positive design parameters with appropriate dimension.

Remark 9.4 *Different from the case with relative degree $\rho_i = 1$, adaptive backstepping technique is employed to deal with the unmatched unknown system parameters in this part. Note that if we follow the standard adaptive backstepping design procedure as in [24], the neighboring states are usually introduced to generate virtual control inputs in the first step. However, for the event-triggered communication case, the received neighboring states in each agent are no longer continuous, which leads to the non-differentiability problem of the virtual control inputs. To overcome this difficulty, an auxiliary system (9.24) with the triggered local consensus error $\bar{\zeta}_i$ being its input is first introduced. Then we replace the triggered local consensus error $\bar{\zeta}_i$ with the differentiable signal $s_{i,1}$ as discussed in Remark 9.3. By doing this, the introduction of neighboring outputs is postponed until the ρ_ith step, which is contained in α_{i,ρ_i}. Since α_{i,ρ_i} is the virtual control input chosen in the final step of local control design, the derivative of α_{i,ρ_i} is not needed. Hence the non-differentiability problem of the virtual control inputs can be solved. On the other hand, some other neighboring states are not involved in agent i, except for the neighboring outputs. Naturally, the system relative degree of all the agents are allowed to be arbitrary.*

9.3 STABILITY AND CONSENSUS ANALYSIS

In this section, the main results of this chapter are formally stated as follows.

Theorem 9.1 *Consider a group of uncertain heterogeneous linear multi-agent systems (9.1) consisting of the decentralized event-triggering conditions (9.6), auxiliary systems (9.24), distributed adaptive control inputs (9.13), (9.37), and a set of parameter estimators (9.15)–(9.19), (9.29)–(9.31), (9.38). Suppose that Assumptions 9.1.1–9.1.3 can be satisfied. For any initial conditions satisfying $V(0) \leq p$, where $V(t)$ is a Lyapunov function defined in (9.39) and p is a given positive constant, there exist design parameters $c_{i,l}$, γ_{D_i}, κ_{D_i}, γ_{ϱ_i}, κ_{ϱ_i}, γ_{P_i}, κ_{P_i}, γ_{y_i}, κ_{y_i}, κ_{θ_i}, Γ_i such that the following conclusions can be drawn.*
(i) All the closed-loop signals are uniformly bounded.
(ii) The vector of actual tracking error $\delta = [\delta_1, \delta_2, ..., \delta_N]^T$ will converge to a compact set.
(iii) The upper bound of $\|\delta(t)\|^2_{[0,T]} = \frac{1}{T}\int_0^T \|\delta(t)\|^2 dt$ can be decreased by choosing suitable design parameters.

Proof. The Lyapunov function for the entire closed-loop system is defined as

$$V = \sum_{i=1}^{N}(1-\nu_i)V_{i,1} + \sum_{i=1}^{N}\nu_i\left(V_{i,\rho_i-1} + \frac{1}{2}z_{i,\rho_i}^2 + \frac{\epsilon_i^T \Phi_i \epsilon_i}{2d_{i,\rho_i}}\right), \qquad (9.39)$$

where $\nu_i = 0$ if the relative degree of agent i is equal to one, otherwise $\nu_i = 1$. From (9.22), (9.36)–(9.38), and Lemma 2.4, the derivative of V is computed as

$$\dot{V} \leq - k\delta^T P(\mathcal{L}+\mathcal{B})\delta - \sum_{i=1}^{N}c_{i,1}z_{i,1}^2 - \sum_{i=1}^{N}\sum_{l=2}^{\rho_i}\nu_i c_{i,l}z_{i,l}^2$$

$$- \sum_{i=1}^{N}\frac{k\kappa_{P_i}}{2}\tilde{P}_i^2 - \sum_{i=1}^{N}\frac{\kappa_{\theta_i}\|\tilde{\theta}_i\|^2}{2} - \sum_{i=1}^{N}\frac{|b_{i,m_i}|\kappa_{\varrho_i}}{2}\tilde{\varrho}_i^2$$

$$- \sum_{i=1}^{N}\frac{(1-\nu_i)\kappa_{D_i}}{2}\tilde{D}_i^2 - \sum_{i=1}^{N}\frac{kP_i(1-\mu_i)\kappa_{y_i}}{4}\tilde{y}_{i,r}^2$$

$$- \sum_{i=1}^{N}\left(\frac{1}{4d_{i,1}} + \sum_{l=2}^{\rho_i}\nu_i\frac{1}{4d_{i,l}}\right)\epsilon_i^T\epsilon_i + \sum_{i=1}^{N}\nu_i M_{i,n}$$

$$- \sum_{i=1}^{N}\sum_{l=1}^{\rho_i-1}\nu_i\tau_{i,0}\tilde{s}_{i,l}^2 + \sum_{i=1}^{N}(1-\nu_i)M_{i,1}$$

$$\leq - \frac{k}{2}\lambda_{\min}(Q)\|\delta\|^2 - \sigma V + M, \qquad (9.40)$$

where $\sigma = \min\left\{2c_{i,l}, 2\tau_{i,0}, \gamma_{D_i}\kappa_{D_i}, \gamma_{\varrho_i}\kappa_{\varrho_i}, \gamma_{P_i}\kappa_{P_i}, \frac{\gamma_{y_i}\kappa_{y_i}}{2}, \frac{\kappa_{\theta_i}}{\lambda_{\max}(\Gamma_i^{-1})}, \frac{1}{2\lambda_{\max}(\Phi_i)}\right\}$

and $M_{i,n} = M_{i,2} + \frac{\kappa_{\theta_i}\|\theta_i-\theta_{i,0}\|^2}{2}$ and $M = \sum_{i=1}^{N}(1-\nu_i)M_{i,1} + \sum_{i=1}^{N}\nu_i M_{i,n}$.

Now we establish the results stated in Theorem 9.1 one by one.

1) We define a compact set Ω as $\Omega = \{(\Omega_1, \ldots, \Omega_N) : V(t) \leq p\}$, where $\Omega_i = \{\epsilon_i, \tilde{\theta}_i, z_{i,1}, \ldots, z_{i,\rho_i}, \nu_i \tilde{s}_{i,1}, \ldots, \nu_i \tilde{s}_{i,\rho_i - 1}, \tilde{\varrho}_i, \tilde{P}_i, (1 - \nu_i)\tilde{D}_i, (1 - \mu_i)\tilde{y}_{i,r}\}$, $i \in \mathcal{V}$. For all the arguments within the compact set Ω, the signals $\epsilon_i, z_{i,1}, \ldots, z_{i,\rho_i}$, $\tilde{s}_{i,1}, \ldots, \tilde{s}_{i,\rho_i - 1}, \hat{\varrho}_i, \hat{P}_i, \hat{\theta}_i, \hat{D}_i$ and $\hat{y}_{i,r}$ are uniformly bounded. In view of the definition of $z_{i,1}$ in (9.8) and Assumption 9.1.2, the system output y_i and the actual tracking error δ_i are bounded. From (9.3), ξ_i, and Ξ_i, it can be concluded that η_i, ξ_i, and Ξ_i are bounded since $A_{i,0}$ is Hurwitz. Then, along the analysis lines in [56], the boundedness of λ_i, x_i, and u_i can also be ensured. Therefore, all the signals in the closed-loop system are uniformly bounded.

From (9.9) and (9.10), it can be concluded that $\delta_i = z_{i,1} - (1 - \mu_i)\tilde{y}_{i,r}$ and $\dot{\zeta}_i = \sum_{j=1}^{N} a_{ij}(\dot{y}_i - \dot{y}_j) + \mu_i(\dot{y}_i - \dot{y}_r)$ are bounded. Hence, the function $\frac{\rho_i - 1}{4}(kP_i\delta_i)^2 + \frac{1}{4}(\tau_{i,\rho_i - 1}\bar{d}_i + |\dot{\zeta}_i|)^2 + \frac{1}{4}(\tau_{i,\rho_i - 1}\bar{d}_i)^2$ contained in $M_{i,2}$ is bounded, so is $M_{i,n}$. We define $\sigma_{i,1} = \min\{2c_{i,1}, \gamma_{D_i}\kappa_{D_i}, \gamma_{\varrho_i}\kappa_{\varrho_i}, \gamma_{P_i}\kappa_{P_i}, \frac{\gamma_{y_i}\kappa_{y_i}}{2}, \frac{\kappa_{\theta_i}}{\lambda_{\max}(\Gamma_i^{-1})}\}$ for the agent i with $\rho_i = 1$ and $\sigma_{i,2} = \min\{2c_{i,2}, \ldots, 2c_{i,\rho_i}\}$ for the agent i with $\rho_i \geq 2$. By choosing the design parameters $c_{i,l}, \gamma_{D_i}, \kappa_{D_i}, \gamma_{\varrho_i}, \kappa_{\varrho_i}, \gamma_{P_i}, \kappa_{P_i}, \gamma_{y_i}, \kappa_{y_i}, \kappa_{\theta_i}, \Gamma_i$ to make $\sigma_{i,1}$ and $\sigma_{i,2}$ to satisfy $\sigma_{i,1} \geq \frac{M_{i,1}}{p}$ and $\sigma_{i,2} \geq \frac{M_{i,n}}{p}$, respectively, we can easily check from (9.40) that $\dot{V}(t) \leq 0$ for $V(t) = p$. Therefore, the compact set Ω is an invariant compact set, which implies that if $V(0) \leq p$, then $V(t) \leq p$ for all $t > 0$.

2) Recalling the definition of $M_{i,1}$, $M_{i,2}$, $M_{i,n}$, and M, there exists a positive constant $M^\star$ such that $M \leq M^\star$. Hence, it follows from (9.40) that

$$\dot{V} \leq -\frac{k}{2}\lambda_{\min}(Q)\|\delta\|^2 - \sigma V + M^\star. \tag{9.41}$$

Neglecting the term $-\frac{k}{2}\lambda_{\min}(Q)\|\delta\|^2$ and integrating both sides of it, we can get $V(t) \leq V(0)e^{-\sigma t} + \frac{M^\star}{\sigma}(1 - e^{-\sigma t}) \leq V(0) + \frac{M^\star}{\sigma}$. From (9.9) and (9.39), we have $\|\delta\|^2 \leq \sum_{i=1}^{N}[2z_{i,1}^2 + 2(1 - \mu_i)^2\tilde{y}_{i,r}^2] \leq \varpi V$, where $\varpi = \max\left\{4, \frac{4(1 - \mu_i)\gamma_{y_i}}{kP_i}\right\}$ for $i \in \mathcal{V}$. Consequently, we can get $\|\delta(t)\|^2 \leq \varpi[V(0)e^{-\sigma t} + \frac{M^\star}{\sigma}(1 - e^{-\sigma t})]$, which implies that the tracking errors in Euclidean norm will converge to a compact set $E_r = \{\delta | \|\delta\|^2 \leq \varpi(M^\star + \varsigma)/\sigma\}$ for $t \geq (1/\sigma)\ln(|V(0)\sigma - M^\star|/\varsigma)$, where ς is an arbitrarily small positive constant.

3) Based on the discussion above, we have $\dot{V} \leq -\frac{k}{2}\lambda_{\min}(Q)\|\delta\|^2 + M^\star$. Integrating both sides of this inequality gets $\|\delta(t)\|_{[0,T]}^2 = \frac{1}{T}\int_0^T \|\delta(t)\|^2 dt \leq \frac{2}{k\lambda_{\min}(Q)}\left[\frac{V(0) - V(T)}{T} + M^\star\right] \leq \frac{2}{k\lambda_{\min}(Q)}\left[\frac{V(0)}{T} + M^\star\right]$. In view of the definition of $M_{i,1}$, $M_{i,2}$, and $M^\star$, it can be concluded that the upper bound of the actual tracking errors δ_i, $i \in \mathcal{V}$, in the mean square sense can be decreased by increasing k, γ_{y_i} and decreasing $\kappa_{\varrho_i}, \kappa_{D_i}, \kappa_{P_i}, \kappa_{\theta_i}, \pi_{i,2}$.

The following theorem shows that Zeno behavior in each agent is ruled out.

Theorem 9.2 *With the same conditions in Theorem 9.1 and the triggering condition (9.6), Zeno behavior in each agent can be avoided. Moreover, the inter-event time internals are lower bounded by a positive constant ς_i, i.e., $\tau^i_{k_i} = t^i_{k_i+1} - t^i_{k_i} \geq \varsigma_i$, $i \in \mathcal{V}, \forall k_i \in Z^+$.*

Proof. Define a set of measurement errors as $e_{y_i}(t) = y_i(t) - y_i(t^i_{k_i}), i \in \mathcal{V}, \forall k_i \in Z^+$. Assume that the triggering condition (9.6) triggers at time instant $t^i_{k_i}$ and $e_{y_i}(t^i_{k_i}) = 0$. During the time interval $[t^i_{k_i}, t^i_{k_i+1})$, the derivative of e_{y_i} is calculated as $\dot{e}_{y_i} = \dot{y}_i = b_{i,m_i} v_{im_i,2} + \xi_{i,2} + \bar{\omega}^T_i \theta_i + \epsilon_{i,2}$. Integrating both sides of this equation gets $e_{y_i}(t) = \int^t_{t^i_{k_i}} b_{i,m_i} v_{im_i,2}(\tau) + \xi_{i,2}(\tau) + \bar{\omega}_i(\tau)^T \theta_i + \epsilon_{i,2}(\tau) d\tau$. Since the boundedness of all the closed-loop signals has been ensured in Theorem 9.1, there exists a positive constant $\bar{e}_{y_i}$ such that $e_{y_i}(t) \leq \bar{e}_{y_i}(t - t^i_{k_i})$. From (9.6), the next event will not trigger before $e_{y_i}(t) = \pi_{i,1}$. Therefore, the inter-event time internals are lower bounded by $\tau^i_{k_i} \geq \frac{\pi_{i,1}}{\bar{e}_{y_i}} \triangleq \varsigma_i, i \in \mathcal{V}, \forall k_i \in Z^+$.

Remark 9.5 *The linear systems (9.1) can be regarded as a special case of the strict-feedback nonlinear systems taking the output-feedback form in [56]. To extend current results to the latter nonlinear systems, the non-differentiability problem as discussed in Remark 9.4 will still exist in the recursive design of distributed adaptive controllers with event-triggered communication. To handle this problem, an auxiliary system of similar form as (9.24) can be introduced in each agent with $\rho_i \geq 2$. Then the virtual control inputs for the first $(\rho_i - 1)$ steps in each agent can be designed by using only local signals, as observed from (9.27)–(9.31) and (9.35). The discrete-time consensus errors $\bar{\zeta}_i$ as defined in (9.11) need only be involved in the ρ_ith design step. Besides, the local filters need be redesigned to generate the state estimates, which can be found in [56].*

9.4 SIMULATION RESULTS

We consider a group of 4 single-link manipulators of the form [120].

$$J_i \ddot{\vartheta}_i + B_i \dot{\vartheta}_i + N_i \sin(\vartheta_i) = \tau_i \text{ for } i = 1, 2, 3, 4. \tag{9.42}$$

where ϑ_i denotes the angle of the link. τ_i denotes the generalized force, which is the control input of manipulators 1 and 3. $J_i = 1\text{kg/m}^2$ is the mechanical inertia. $B_i = 0.1\text{Nms/rad}$ is the coefficient of viscous friction at the joint. $N_i = 5\text{N} \cdot \text{m}$ is a positive constant related to the mass of the load and the coefficient of gravity. For manipulators 2 and 4, the motor dynamics modeled as follows are also considered.

$$Y_i \dot{\tau}_i + R_i \tau_i = u_i - K_i \dot{\vartheta}_i \text{ for } i = 2, 4. \tag{9.43}$$

where $Y_i = 1\text{H}$, $R_i = 1\Omega$, and $K_i = 0.3\text{Nm/A}$ are the armature inductance, armature resistances, and back electromotive force coefficient, respectively. u_i is the voltage applied to the motor, which is the control input of manipulators 2 and 4. Therefore, this group of manipulators consist of two second-order systems with relative degree 2 and two third-order systems with relative degree 3.

Define $x_{i,1} = \vartheta_i$, $x_{i,2} = \dot{\vartheta}_i$, and $x_{i,3} = \tau_i$. By linearizing manipulator i for $i = 1, 2, 3, 4$ at origin, we have

$$\begin{bmatrix} \dot{x}_{i,1} \\ \dot{x}_{i,2} \end{bmatrix} = \begin{bmatrix} 0 & 1 \\ -\frac{N_i}{J_i} & -\frac{B_i}{J_i} \end{bmatrix} \begin{bmatrix} x_{i,1} \\ x_{i,2} \end{bmatrix} + \begin{bmatrix} 0 \\ \frac{1}{J_i} \end{bmatrix} \tau_i, \text{ for } i = 1, 3, \qquad (9.44)$$

and

$$\begin{bmatrix} \dot{x}_{i,1} \\ \dot{x}_{i,2} \\ \dot{x}_{i,3} \end{bmatrix} = \begin{bmatrix} 0 & 1 & 0 \\ -\frac{N_i}{J_i} & -\frac{B_i}{J_i} & \frac{1}{J_i} \\ 0 & -\frac{K_i}{Y_i} & -\frac{R_i}{Y_i} \end{bmatrix} \begin{bmatrix} x_{i,1} \\ x_{i,2} \\ x_{i,3} \end{bmatrix} + \begin{bmatrix} 0 \\ 0 \\ \frac{1}{Y_i} \end{bmatrix} u_i, \text{ for } i = 2, 4. \qquad (9.45)$$

Following the similar manipulation as conducted in [56, Page 313], systems (9.44) and (9.45) can be transformed into the form of system (9.1). For manipulators 1 and 3, the system parameters are $b_{i,0} = \frac{1}{J_i}$, $\phi_{i,1} = \frac{B_i}{J_i}$ and $\phi_{i,0} = \frac{N_i}{J_i}$ for $i = 1, 3$. For manipulators 2 and 4, the system parameters are $b_{i,0} = \frac{1}{J_i Y_i}$, $\phi_{i,2} = \frac{B_i}{J_i} + \frac{R_i}{Y_i}$, $\phi_{i,1} = \frac{N_i}{J_i} + \frac{K_i}{J_i Y_i} + \frac{B_i R_i}{J_i Y_i}$, and $\phi_{i,0} = \frac{N_i R_i}{J_i Y_i}$ for $i = 2, 4$. It can be checked that the resulted linear systems in the form of system (9.1) satisfy Assumption 9.1.3.

The desired trajectory is represented by $y_r(t) = 0.5 \sin(0.1t) + 0.5 \sin(0.05t)$. The graph among the 4 manipulators is depicted in Figure 9.1, where only manipulator 1 has access to the desired trajectory $y_r(t)$.

In simulation, all the initial states are set as zero except that $x_{1,1}(0) = 0.8$, $x_{2,1}(0) = 0.7$, $x_{3,1}(0) = 0.5$, $x_{4,1}(0) = 0.6$, $s_{1,1}(0) = 0.8$, $s_{2,1}(0) = s_{2,2}(0) = 0.1$, $s_{3,1}(0) = -0.2$, $s_{4,1}(0) = s_{4,2}(0) = -0.2$. The design parameters are chosen

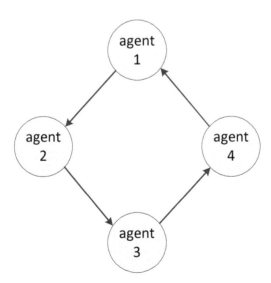

FIGURE 9.1 The graph among the 4 manipulators.

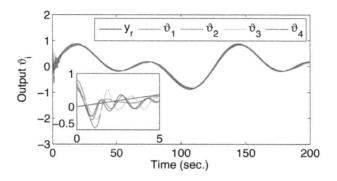

FIGURE 9.2   The outputs y_i, $i = 1, \ldots, 4$.

as $\pi_{i,1} = 0.02$, $\pi_{i,2} = 0.2$, $M_i = 0.2$, $k_{i,1} = 1$, $k_{i,2} = 2$, $k_{2,3} = k_{4,3} = 1$, $\tau_{i,0} = 1$, $k = 1$, $d_{i,1} = d_{i,2} = d_{i,3} = 0.1$, $c_{i,1} = c_{i,2} = 1$, $c_{i,3} = 5$, $\varepsilon_i = 0.06$, $\gamma_{\varrho_i} = \gamma_{P_i} = \gamma_{D_i} = 1$, $\gamma_{y_i} = 5$, $\kappa_{\varrho_i} = \kappa_{P_i} = \kappa_{D_i} = \kappa_{y_i} = \kappa_{\theta_i} = 0.005$, $\varrho_{i,0} = P_{i,0} = D_{i,0} = y_{i,0} = 0.005$. Γ_i and $\theta_{i,0}$ are chosen as identity matrices and zero matrices with appropriate dimension, respectively. The tracking performance of each agent is shown in Figures 9.2–9.3, which shows that the output of each agent can track the desired trajectory $y_r(t)$ with a bounded tracking error. Figure 9.4 exhibits the control inputs of all agents. The triggering times are shown in Figure 9.5. It can be seen that all the observed signals are bounded while Zeno behavior in each agent is ruled out.

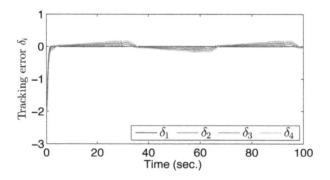

FIGURE 9.3   The tracking errors $\delta_i = y_i - y_r$, $i = 1, \ldots, 4$.

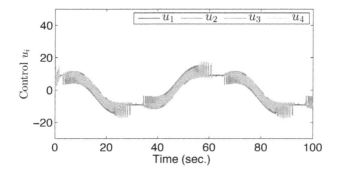

FIGURE 9.4   The controllers u_i, $i = 1, \ldots, 4$.

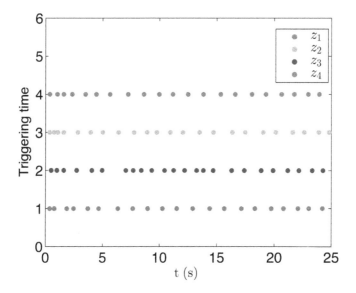

FIGURE 9.5   The triggering times of y_i, $i = 1, \ldots, 4$.

9.5   NOTES

In this chapter, a novel distributed adaptive control scheme is proposed for heterogenous linear multi-agent systems with unknown system parameters, event-triggered communication and directed graph via output feedback control. It is shown that with the proposed control scheme, all the closed-loop signals are uniformly bounded and the desired output consensus tracking can be achieved. Besides, Zeno behavior in each agent is ruled out. Finally, the effectiveness of the proposed control scheme has been verified by an illustrative example.

Section IV

Applications

10 Distributed Adaptive Formation Control of Nonholonomic Mobile Robots

In this chapter, we apply the distributed adaptive tracking control strategy in Chapter 3 to solve the formation control problem for multiple nonholonomic mobile robots with unknown parameters. Note that the models of the mobile robots are a type of uncertain underactuated mechanical systems, which brings new difficulties in designing distributed adaptive controllers. Therefore, only a few results have been reported in this area so far and the system models considered in most of the existing results are confined to kinematic model. Motivated by these, we investigate the formation control problem for multiple uncertain nonholonomic mobile robots with both kinematic and dynamic models under the condition that only part of the robots can access the exact information of the reference directly. To solve the underactuation in the kinematic level, the transverse function approach is applied to create an additional virtual control, which is free to design. It is proved that the formation errors of the overall system can be made as small as desired by adjusting the design parameters properly with the combination of our proposed distributed control strategy and the transverse function technique in [84].

10.1 PROBLEM FORMULATION

10.1.1 SYSTEM MODEL

We consider a group of N two-wheeled mobile robots, shown in Figure 10.1, each of which can be described by the following dynamic model [25].

$$\dot{\eta}_i = J(\eta_i)\omega_i \tag{10.1}$$

$$M_i\dot{\omega}_i + C_i(\dot{\eta}_i)\omega_i + D_i\omega_i = \tau_i, \ \text{for } i = 1, \ldots, N \tag{10.2}$$

where $\eta_i = [\bar{x}_i, \bar{y}_i, \bar{\phi}_i]^T$ denotes the position and orientation of the ith robot. $\omega_i = [\omega_{i1}, \omega_{i2}]^T$ denotes the angular velocities of the left and right wheels, and $\tau_i = [\tau_{i1}, \tau_{i2}]^T$ represents the control torques applied to the wheels. M_i is a symmetric, positive definite inertia matrix, $C_i(\dot{\eta}_i)$ is the centripetal and Coriolis matrix, and D_i denotes the surface friction. These matrices have the same form as

DOI: 10.1201/9781003394372-10

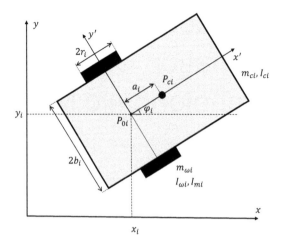

FIGURE 10.1   The nonholonomic mobile robot.

those in [25], which are given below for completeness.

$$J(\eta_i) = \frac{r_i}{2} \begin{bmatrix} \cos\bar{\phi}_i & \cos\bar{\phi}_i \\ \sin\bar{\phi}_i & \sin\bar{\phi}_i \\ b_i^{-1} & -b_i^{-1} \end{bmatrix}, \quad M_i = \begin{bmatrix} m_{i1} & m_{i2} \\ m_{i2} & m_{i1} \end{bmatrix}$$

$$C_i(\dot{\eta}_i) = \begin{bmatrix} 0 & c_i\dot{\bar{\phi}}_i \\ -c_i\dot{\bar{\phi}}_i & 0 \end{bmatrix}, \quad D_i = \begin{bmatrix} d_{i1} & 0 \\ 0 & d_{i2} \end{bmatrix}$$

$$m_{i1} = \frac{1}{4}b_i^{-2}r_i^2(m_ib_i^2 + I_i) + I_{wi}$$

$$m_{i2} = \frac{1}{4}b_i^{-2}r_i^2(m_ib_i^2 - I_i)$$

$$I_i = m_{ci}l_i^2 + 2m_{wi}b_i^2 + I_{ci} + 2I_{mi}$$

$$c_i = \frac{1}{2}b_i^{-1}r_i^2 m_{ci}l_i, \quad m_i = m_{ci} + 2m_{wi}. \tag{10.3}$$

In (10.3), m_{ci}, m_{wi}, I_{ci}, I_{wi}, I_{mi}, and d_{ik} are unknown system parameters of which the physical meanings can be found in [25].

Remark 10.1 *Observing (10.1) with $J(\eta_i)$ in (10.3), it can be seen that the number of inputs (i.e., ω_{i1} and ω_{i2}) is less than the number of configuration variables (i.e., $\bar{x}_i$, $\bar{y}_i$, and $\bar{\phi}_i$). Thus the considered mobile robot is an underactuated mechanical system. To achieve the tracking objectives of $\bar{x}_i$, $\bar{y}_i$, and $\bar{\phi}_i$ separately, transverse function approach [84] will be employed. An auxiliary manipulated variable will be introduced with which the underactuated problem can be transformed to a fully-actuated one.*

10.1.2 CHANGE OF COORDINATES

We change the original coordinates of the ith robot as follows:

$$\begin{bmatrix} x_i \\ y_i \end{bmatrix} = \begin{bmatrix} \bar{x}_i \\ \bar{y}_i \end{bmatrix} + R(\phi_i) \begin{bmatrix} f_{1i}(\xi_i) \\ f_{2i}(\xi_i) \end{bmatrix} \qquad (10.4)$$

$$\phi_i = \bar{\phi}_i - f_{3i}(\xi_i), \qquad (10.5)$$

where

$$R(\phi_i) = \begin{bmatrix} \cos(\phi_i) & -\sin(\phi_i) \\ \sin(\phi_i) & \cos(\phi_i) \end{bmatrix} \qquad (10.6)$$

and $f_{li}(\xi_i)$ for $l = 1, 2, 3$ are functions of ξ_i designed as

$$f_{1i}(\xi_i) = \varepsilon_{1i} \sin(\xi_i) \frac{\sin(f_{3i})}{f_{3i}}$$

$$f_{2i}(\xi_i) = \varepsilon_{1i} \sin(\xi_i) \frac{1 - \cos(f_{3i})}{f_{3i}}$$

$$f_{3i}(\xi_i) = \varepsilon_{2i} \cos(\xi_i) \qquad (10.7)$$

with ε_{1i} and ε_{2i} being positive constants and ε_{2i} satisfying $0 < \varepsilon_{2i} < \frac{\pi}{2}$. The following properties can be easily shown.

$$|f_{1i}| < \varepsilon_{1i}, |f_{2i}| < \varepsilon_{1i}, |f_{3i}| < \varepsilon_{2i}. \qquad (10.8)$$

Computing the derivatives of x_i, y_i, and ϕ_i yields that

$$\begin{bmatrix} \dot{x}_i \\ \dot{y}_i \end{bmatrix} = Q_i \begin{bmatrix} r_i u_{i1} \\ \dot{\xi}_i \end{bmatrix} + \frac{\partial R(\phi_i)}{\partial \phi_i} \begin{bmatrix} f_{1i}(\xi_i) \\ f_{2i}(\xi_i) \end{bmatrix}$$

$$\times \left(r_i b_i^{-1} u_{i2} - \frac{\partial f_{3i}(\xi_i)}{\partial \xi_i} \dot{\xi}_i \right) \qquad (10.9)$$

$$\dot{\phi}_i = r_i b_i^{-1} u_{i2} - \frac{\partial f_{3i}(\xi_i)}{\partial \xi_i} \dot{\xi}_i, \qquad (10.10)$$

where $u_{i1} = 0.5(\omega_{i1} + \omega_{i2})$, $u_{i2} = 0.5(\omega_{i1} - \omega_{i2})$ and

$$Q_i = \left[\begin{pmatrix} \cos(\bar{\phi}_i) \\ \sin(\bar{\phi}_i) \end{pmatrix} \quad R(\phi_i) \begin{pmatrix} \frac{\partial f_{1i}(\xi_i)}{\partial \xi_i} \\ \frac{\partial f_{2i}(\xi_i)}{\partial \xi_i} \end{pmatrix} \right] \qquad (10.11)$$

is ensured to be invertible [84]. Different from $(\bar{x}_i, \bar{y}_i, \bar{\phi}_i)$, the transformed coordinates (x_i, y_i, and ϕ_i) can be controlled separately by tuning u_{i1}, u_{i2}, and $\dot{\xi}_i$ is deemed as an auxiliary manipulated variable.

10.1.3 FORMATION CONTROL OBJECTIVE

The components of the desired trajectory in X and Y directions can be expressed respectively as

$$x_r(t) = w_r f_{rx}(t) + c_{rx} \quad \text{and} \quad y_r(t) = w_r f_{ry}(t) + c_{ry}. \tag{10.12}$$

Similar to Chapter 4, it is assumed that $f_{rx}(t)$ and $f_{ry}(t)$ are known by all the robots, whereas the parameters w_r, c_{rx}, and c_{ry} are only available to part of the robots. Besides, $\phi_r(t) \triangleq \arctan\left(\frac{\dot{y}_r}{\dot{x}_r}\right)$ denotes the reference trajectory for the orientation of each robot.

The *control objective* in this chapter is to design distributed adaptive formation controllers such that all the robots can follow a desired trajectory in X-Y plane by maintaining certain prescribed demanding distances from the desired trajectory, i.e.,

$$\lim_{t \to \infty} [x_i(t) - x_r(t)] = -\rho_{ix} \tag{10.13}$$

$$\lim_{t \to \infty} [y_i(t) - y_r(t)] = -\rho_{iy} \tag{10.14}$$

$$\lim_{t \to \infty} [\phi_i(t) - \phi_r(t)] = 0. \tag{10.15}$$

Similar to Chapter 4, we suppose that the communication status among the N robots can be represented by a directed graph $\mathcal{G}$ and Assumption 3.1.3 holds. To achieve the formation control objective, the following assumptions are also needed.

Assumption 10.1.1 f_{rx}, f_{ry}, $\dot{f}_{rx}$, $\dot{f}_{ry}$ and $\ddot{f}_{rx}$, $\ddot{f}_{ry}$ are bounded, piece-wise continuous bounded and known to all the robots.

Assumption 10.1.2 *The parameters r_i and b_i fall in known compact sets, i.e., there exist some known positive constants $\bar{r}_i$, $\underline{r}_i$, $\bar{b}_i$, and $\underline{b}_i$ such that $\underline{r}_i < r_i < \bar{r}_i$ and $\underline{b}_i < b_i < \bar{b}_i$.*

Assumption 10.1.3 *The demanding distances ρ_{ix} and ρ_{iy} for robot i are available to its neighbors.*

Remark 10.2

1. *It can be seen that the consensus tracking objective (ii) stated in Chapter 3, i.e., $\lim_{t \to \infty}[y_i(t) - y_r(t)] = 0$, is actually a special case of the formation objectives in (10.13)–(10.15) with $\rho_{ix} = 0$ and $\rho_{iy} = 0$. In contrast to the fact that exact information about $x_r(t)$ and $y_r(t)$ are only accessible to a subset of the robots, the desired orientation $\phi_r(t) = \arctan\left(\frac{\dot{y}_r}{\dot{x}_r}\right)$ is available to all the robots since $f_{rx}(t)$ and $f_{ry}(t)$ are available to all the robots.*
2. *Note that (10.9), (10.10), and (10.2) constitute the new system to be controlled. In (10.9)–(10.10), u_{i1}, $\dot{\xi}_i$ and u_{i2} act as the control inputs while x_i, y_i, and ϕ_i are the outputs. Thus different from the traditional underactuated kinematic model*

for mobile robots, the new multi-input multi-output (MIMO) kinematic model can be treated as three separate single-input single-output (SISO) systems with the aid of transverse function technique. Moreover, since τ_{i1} and τ_{i2} in (10.2) are the actual control inputs of each robot system, the relative degree of the entire system at dynamic model level is two. This indicates that the backstepping-based adaptive control scheme proposed for one-dimensional output consensus tracking problem in Chapter 3 can be extended to solve the formation control problem in this chapter.

3. *From (10.4), (10.5), and the properties of f_{li} in (10.8), it is clear that the transformation errors $x_i - \bar{x}_i$, $y_i - \bar{y}_i$, $\phi_i - \bar{\phi}_i$ are bounded by ε_{1i} and ε_{2i}. It will be shown that the designed distributed adaptive controllers can guarantee the convergence of the formation control errors with respect to x_i, y_i, and ϕ_i. Therefore, the formation control errors with respect to the true position and orientation, i.e., $\bar{x}_i$, $\bar{y}_i$, and $\bar{\phi}_i$, can be made as small as desired by adjusting ε_{1i} and ε_{2i} properly.*

10.2 DISTRIBUTED ADAPTIVE FORMATION CONTROLLER DESIGN

As discussed in Remark 10.2, the control design procedure in this part involves two steps by adopting the backstepping technique. In the first step, the virtual controls for u_{i1}, u_{i2} and the auxiliary manipulated variable $\dot{\xi}_i$ will be chosen. In the second step, the actual control inputs τ_i will be derived.

Step 1. Define local error variables as

$$z_{ix,1} = \sum_{j=1}^{N} a_{ij}(x_i + \rho_{ix} - x_j - \rho_{jx}) + \mu_i(x_i + \rho_{ix} - x_r)$$

$$z_{iy,1} = \sum_{j=1}^{N} a_{ij}(y_i + \rho_{iy} - y_j - \rho_{jy}) + \mu_i(y_i + \rho_{iy} - y_r)$$

$$e_{ix,1} = x_i - \mu_i x_r - (1 - \mu_i)\left(f_{rx}\hat{w}_{rx,i} - \hat{c}_{rx,i}\right) + \rho_{ix}$$

$$e_{iy,1} = y_i - \mu_i y_r - (1 - \mu_i)\left(f_{ry}\hat{w}_{ry,i} - \hat{c}_{ry,i}\right) + \rho_{iy}$$

$$\delta_{i\phi} = \phi_i - \phi_r$$

$$e_{ix,2} = u_{i1} - \alpha_{i1}, \ e_{i\phi,2} = u_{i2} - \alpha_{i2} \tag{10.16}$$

where $\hat{w}_{rx,i}$ ($\hat{w}_{ry,i}$), $\hat{c}_{rx,i}$, and $\hat{c}_{ry,i}$ are the estimates introduced in the ith robot for the unknown trajectory parameters if $\mu_i = 0$. We choose the virtual controls $(\alpha_{i1}, \alpha_{i2})$ and $\dot{\xi}$ in transverse function technique as

$$\begin{bmatrix} \alpha_{i1} \\ \dot{\xi}_i \end{bmatrix} = \begin{bmatrix} \hat{\theta}_{i1}^{-1} & 0 \\ 0 & 1 \end{bmatrix} Q_i^{-1}\Omega_i \tag{10.17}$$

$$\alpha_{i2} = \hat{\theta}_{i2}^{-1}\left(-k_2\delta_{i\phi} + \frac{\partial f_{3i}(\xi_i)}{\partial \xi_i}\dot{\xi}_i + \dot{\phi}_r\right), \tag{10.18}$$

where $\hat{\theta}_{i1}$ and $\hat{\theta}_{i2}$ are the estimates of r_i and $r_i b_i^{-1}$, respectively.

$$
\begin{aligned}
\Omega_i = & -k_1 P_i \begin{bmatrix} z_{ix,1} \\ z_{iy,1} \end{bmatrix} - \frac{\partial R(\phi_i)}{\partial \phi_i} \begin{bmatrix} f_{1i}(\xi_i) \\ f_{2i}(\xi_i) \end{bmatrix} \left(-k_2 \delta_{i\phi} + \dot{\phi}_r \right) \\
& - \begin{bmatrix} \dot{\rho}_{ix} \\ \dot{\rho}_{iy} \end{bmatrix} + \mu_i \begin{bmatrix} \dot{f}_{rx} w_{rx} \\ \dot{f}_{ry} w_{ry} \end{bmatrix} \\
& + (1 - \mu_i) \begin{bmatrix} \dot{f}_{rx} \hat{w}_{rx,i} + f_{rx} \dot{\hat{w}}_{rx,i} + \dot{\hat{c}}_{rx,i} \\ \dot{f}_{ry} \hat{w}_{ry,i} + f_{ry} \dot{\hat{w}}_{ry,i} + \dot{\hat{c}}_{ry,i} \end{bmatrix},
\end{aligned}
\tag{10.19}
$$

where k_1, k_2 being positive constants. P_i is defined in Lemma 2.4. The above design delivers the following results.

$$
\begin{aligned}
\begin{bmatrix} \dot{e}_{ix,1} \\ \dot{e}_{iy,1} \end{bmatrix} = & -k_1 P_i \begin{bmatrix} z_{ix,1} \\ z_{iy,1} \end{bmatrix} + \frac{\partial R(\phi_i)}{\partial \phi_i} \begin{bmatrix} f_{1i}(\xi_i) \\ f_{2i}(\xi_i) \end{bmatrix} \left(\tilde{\theta}_{i2} u_{i2} \right. \\
& \left. + \hat{\theta}_{i2} e_{i\phi,2} \right) + Q_i \begin{bmatrix} \tilde{\theta}_{i1} u_{i1} + \hat{\theta}_{i1} e_{ix,2} \\ 0 \end{bmatrix} \\
\dot{\delta}_{i\phi} = & -k_2 \delta_{i\phi} + \tilde{\theta}_{i2} u_{i2} + \hat{\theta}_{i2} e_{i\phi,2}.
\end{aligned}
\tag{10.20}
$$

The parameter estimators at this step are designed as

$$
\begin{aligned}
& \dot{\hat{w}}_{rx,i} = -\gamma_{ri} f_{rx} e_{ix,1}, \quad \dot{\hat{w}}_{ry,i} = -\gamma_{ri} f_{ry} e_{iy,1} \\
& \dot{\hat{c}}_{rx,i} = -\gamma_{ri} e_{ix,1}, \quad \dot{\hat{c}}_{ry,i} = -\gamma_{ri} e_{iy,1} \\
& \dot{\hat{\theta}}_{i1} = \text{Proj}\left(\hat{\theta}_{i1}, \gamma_{\theta_{i1}} \pi_{i1} u_{i1} \right) \\
& \dot{\hat{\theta}}_{i2} = \text{Proj}\left(\hat{\theta}_{i2}, \gamma_{\theta_{i1}} \pi_{i2} u_{i2} \right)
\end{aligned}
\tag{10.21}
$$

with

$$
\begin{aligned}
\pi_{i1} = & \, e_{ix,1} \cos(\bar{\phi}_i) + e_{iy,1} \sin(\bar{\phi}_i) \\
\pi_{i2} = & \, [e_{ix,1}, e_{iy,1}] \frac{\partial R(\phi_i)}{\partial \phi_i} \begin{bmatrix} f_{1i}(\xi_i) \\ f_{2i}(\xi_i) \end{bmatrix} + \delta_{i\phi}.
\end{aligned}
\tag{10.22}
$$

Note $\text{Proj}(\cdot, \cdot)$ denotes a Lipschitz continuous projection operator about which the design details and properties can be found in [56]. It is adopted here to ensure that $\hat{\theta}_{i1} > 0$ and $\hat{\theta}_{i2} > 0$. Thus $\hat{\theta}_{i1}^{-1}$ and $\hat{\theta}_{i2}^{-1}$ in (10.17) and (10.18) are well defined. We choose a Lyapunov function candidate at this step as

$$
\begin{aligned}
V_1 = & \frac{1}{2} \sum_{i=1}^{N} \left(e_{ix,1}^2 + e_{iy,1}^2 + \delta_{i\phi}^2 + \frac{1}{\gamma_{\theta_{i1}}} \tilde{\theta}_{i1}^2 + \frac{1}{\gamma_{\theta_{i2}}} \tilde{\theta}_{i2}^2 \right) \\
& + \frac{k_1}{2} \sum_{i=1}^{N} (1 - \mu_i) \frac{P_i}{\gamma_{ri}} \left(\tilde{w}_{rx,i}^2 + \tilde{w}_{ry,i}^2 + \tilde{c}_{rx,i}^2 + \tilde{c}_{ry,i}^2 \right)
\end{aligned}
\tag{10.23}
$$

From (10.20) and (10.21), the derivative of V_1 in (10.23) can be computed as

$$
\dot{V}_1 \leq -\frac{k_1}{2}\left(\delta_x^T Q \delta_x + \delta_y^T Q \delta_y\right) - k_2 \delta_{i\phi}^T \delta_{i\phi}
$$

$$
+ \sum_{i=1}^{N}\left(\pi_{i1}\hat{\theta}_{i1}e_{ix,2} + \pi_{i2}\hat{\theta}_{i2}e_{i\phi,2}\right) \tag{10.24}
$$

where $\delta_x = [\delta_{1x}, \ldots, \delta_{Nx}]$ with $\delta_{ix} = x_i - x_r + \rho_{ix}$ and $\delta_y = [\delta_{1y}, \ldots, \delta_{Ny}]$ with $\delta_{iy} = y_i - y_r + \rho_{iy}$. Q is defined in (2.4). The property of projection that $\tilde{b}\mathrm{Proj}(\hat{b}, a) \geq \tilde{b}a$ for $\tilde{b} = b - \hat{b}$ [56] has been used.

Step 2. We now at the position to derive the actual control torque τ_i. Define $\omega_{i1d} = \alpha_{i1} + \alpha_{i2}$, $\omega_{i2d} = \alpha_{i1} - \alpha_{i2}$ and $z_{i,1} = \omega_{i1} - \omega_{i1d}$, $z_{i,2} = \omega_{i2} - \omega_{i2d}$. From (10.16) and the fact that $\omega_{i1} = u_{i1} + u_{i2}$ and $\omega_{i2} = u_{i1} - u_{i2}$, there exist $e_{ix,2} = 0.5(z_{i,1} + z_{i,2})$ and $e_{i\phi,2} = 0.5(z_{i,1} - z_{i,2})$. Let $z_i = [z_{i,1}, z_{i,2}]^T$. Thus we have

$$
z_i = \omega_i - \begin{bmatrix} \omega_{i1d} \\ \omega_{i2d} \end{bmatrix}. \tag{10.25}
$$

Multiplying the derivatives of both sides of (10.25) by M_i and combining it with (10.17) and (10.18), we obtain that

$$
M_i \dot{z}_i = -D_i z_i + \Phi_i^T \Theta_i + \tau_i, \tag{10.26}
$$

where matrix Φ_i and Θ_i are defined as

$$
\Phi_i = [\chi_i, \chi_{i,j_1}, \chi_{i,j_2}, \ldots, \chi_{i,j_{n_i}}]^T, \tag{10.27}
$$

$$
\Theta_i = [\vartheta_i^T, \vartheta_{i,j_1}^T, \vartheta_{i,j_2}^T, \ldots, \vartheta_{i,j_{n_i}}^T]^T. \tag{10.28}
$$

Note j_p for $p = 1, \ldots, n_i$ are the indexes of robot i's neighboring robots (i.e., $j_p \in \mathcal{N}_i$) of which the total number is n_i. The elements in Φ_i and Θ_i are given in (10.31).

Remark 10.3 Θ_i in (10.28) is a vector of unknown parameters involved in the ith robot dynamic subsystem. ϑ_i is the local unknown parameter, while ϑ_{i,j_p} is the coupled uncertainty related to the unknown parameters in robot j_p's dynamics if $a_{ij_p} = 1$. Thus online estimates of ϑ_{i,j_p}, i.e., $\hat{\vartheta}_{i,j_p}$, will be introduced in designing the torques for robot i.

Introduce the estimate $\hat{\Theta}_i$ for unknown parameter vector Θ_i. Then the local control torque and adaptive law are designed as

$$
\tau_i = -K_i z_i - \Phi_i^T \hat{\Theta}_i - 0.5\Xi_i, \tag{10.29}
$$

$$
\dot{\hat{\Theta}}_i = \Gamma_i \Phi_i z_i, \tag{10.30}
$$

$$\vartheta_i = [c_i r_i b_i^{-1} \ d_{i1} \ d_{i2} \ m_{i1} \ m_{i2} \ m_{i1} r_i \ m_{i2} r_i \ m_{i1} r_i b_i^{-1} \ m_{i2} r_i b_i^{-1}]^T,$$

$$\vartheta_{i,jni} = [m_{i1} r_j \ m_{i2} r_j \ m_{i1} r_j b_j^{-1} \ m_{i2} r_j b_j^{-1}]^T$$

$$\chi_i = \begin{bmatrix} -\omega_{i2} u_{i2} & -\omega_{i1d} & 0 & -\Delta_{i11} & -\Delta_{i12} & -\Delta_{i21} & -\Delta_{i22} & -\Delta_{i31} & -\Delta_{i32} \\ \omega_{i1} u_{i2} & 0 & -\omega_{i2d} & -\Delta_{i12} & -\Delta_{i11} & -\Delta_{i22} & -\Delta_{i21} & -\Delta_{i32} & -\Delta_{i31} \end{bmatrix},$$

$$\chi_{ij} = \begin{bmatrix} -\Delta_{ij11} & -\Delta_{ij12} & -\Delta_{ij21} & -\Delta_{ij22} \\ -\Delta_{ij12} & -\Delta_{ij11} & -\Delta_{ij22} & -\Delta_{ij21} \end{bmatrix}$$

$$\Delta_{i1k} = \frac{\partial \omega_{ikd}}{\partial \rho_{ix}} \dot{\rho}_{ix} + \frac{\partial \omega_{ikd}}{\partial \dot{\rho}_{ix}} \ddot{\rho}_{ix} + \frac{\partial \omega_{ikd}}{\partial \rho_{iy}} \dot{\rho}_{iy} + \frac{\partial \omega_{ikd}}{\partial \dot{\rho}_{iy}} \ddot{\rho}_{iy} + \frac{\partial \omega_{ikd}}{\partial \dot{f}_{rx}} \ddot{f}_{rx} + \frac{\partial \omega_{ikd}}{\partial \dot{f}_{ry}} \ddot{f}_{ry}$$

$$+ \frac{\partial \omega_{ikd}}{\partial \dot{\phi}_r} \ddot{\phi}_r + \frac{\partial \omega_{ikd}}{\partial \hat{\theta}_{i1}} \dot{\hat{\theta}}_{i1} + \frac{\partial \omega_{ikd}}{\partial \hat{\theta}_{i2}} \dot{\hat{\theta}}_{i2} + \frac{\partial \omega_{ikd}}{\partial \hat{w}_{rx,i}} \dot{\hat{w}}_{rx,i} + \frac{\partial \omega_{ikd}}{\partial \hat{w}_{ry,i}} \dot{\hat{w}}_{ry,i}$$

$$\Delta_{i2k} = \frac{\partial \omega_{ikd}}{\partial \bar{x}_i}(\cos(\bar{\phi}_i) u_{i1}) + \frac{\partial \omega_{ikd}}{\partial \bar{y}_i}(\sin(\bar{\phi}_i) u_{i1}), \quad \Delta_{i3k} = \frac{\partial \omega_{ikd}}{\partial \bar{\phi}_i} u_{i2}$$

$$\Delta_{ij1k} = \frac{\partial \omega_{ikd}}{\partial \bar{x}_j}(\cos(\bar{\phi}_j) u_{j1}) + \frac{\partial \omega_{ikd}}{\partial \bar{y}_j}(\sin(\bar{\phi}_j) u_{j1}), \quad \Delta_{ij2k} = \frac{\partial \omega_{ikd}}{\partial \bar{\phi}_j} u_{j2}, \quad \text{for } k = 1, 2$$

$$\tag{10.31}$$

---

where $\Xi_i = [\Xi_{i,1}, \Xi_{i,2}]^T$ with

$$\Xi_{i,1} = \pi_{i1} \hat{\theta}_{i1} + \pi_{i2} \hat{\theta}_{i2}, \quad \Xi_{i,2} = \pi_{i1} \hat{\theta}_{i1} - \pi_{i2} \hat{\theta}_{i2}. \tag{10.32}$$

Choose the Lyapunov function for the overall system as

$$V_2 = V_1 + \frac{1}{2}\left(z_i^T M_i z_i + \tilde{\Theta}_i^T \Gamma_i^{-1} \tilde{\Theta}_i\right), \tag{10.33}$$

where Γ_i is a symmetric and positive definite matrix and $\tilde{\Theta}_i = \Theta_i - \hat{\Theta}_i$. From (10.24), (10.29), and (10.30), we obtain that

$$\dot{V}_2(t) \leq -\frac{k_1}{2}\left(\delta_x^T Q \delta_x + \delta_y^T Q \delta_y\right) - k_2 \delta_{i\phi}^T \delta_{i\phi}$$
$$- z_i^T (K_i + D_i) z_i. \tag{10.34}$$

10.3 STABILITY AND CONSENSUS ANALYSIS

The main results in this chapter are formally presented in the following theorem.

Theorem 10.1 *Consider the closed-loop adaptive system consisting of N non-holonomic mobile robots (10.1)–(10.2), the control torques (10.29) and parameter* *estimators (10.21) and (10.30) under Assumptions 3.1.3, 10.1.1–10.1.3. The formation errors for each robot are ensured to satisfy that*

$$\lim_{t\to\infty} \bar{x}_i(t) + \rho_{ix} - x_r(t) \le \sqrt{2}\varepsilon_{i1} \tag{10.35}$$

$$\lim_{t\to\infty} \bar{y}_i(t) + \rho_{iy} - y_r(t) \le \sqrt{2}\varepsilon_{i1} \tag{10.36}$$

$$\lim_{t\to\infty} \bar{\phi}_i(t) - \phi_r(t) \le \varepsilon_{i2}. \tag{10.37}$$

Proof. By following similar analysis to the proof of Theorem 1 and from (10.34), it can be shown that δ_{ix}, δ_{iy}, and $\delta_{i\phi}$ will converge to zero asymptotically. This indicates that $\lim_{t\to\infty}[x_i(t) - x_r(t)] = -\rho_{ix}$, $\lim_{t\to\infty}[y_i(t) - y_r(t)] = -\rho_{iy}$ and $\lim_{t\to\infty}[\phi_i(t) - \phi_r(t)] = 0$.

From (10.4), (10.5), and (10.7), we obtain that

$$\|(x_i - \bar{x}_i, y_i - \bar{y}_i)\| \le \sqrt{2\varepsilon_{i1}^2}, \ |\phi_i - \bar{\phi}_i| \le \varepsilon_{i2}. \tag{10.38}$$

It then follows that

$$\begin{aligned} |\bar{x}_i + \rho_{ix} - x_r| &\le |\bar{x}_i - x_i| + |x_i + \rho_{ix} - x_r| \\ |\bar{y}_i + \rho_{iy} - y_r| &\le |\bar{y}_i - y_i| + |y_i + \rho_{iy} - y_r| \\ |\bar{\phi}_i - \phi_r| &\le |\bar{\phi}_i - \phi_i| + |\phi_i - \phi_r|. \end{aligned} \tag{10.39}$$

Since $x_i + \rho_{ix} - x_r$, $y_i + \rho_{iy} - y_r$, and $\phi_i - \phi_r$ will converge to zero asymptotically, (10.35)–(10.37) hold. As discussed in Remark 10.2, by properly adjusting ε_{i1} and ε_{i2}, the formation errors of the overall system can be made as small as desired.

10.4 SIMULATION RESULTS

Similar to Chapter 3, we now use four mobile robots to demonstrate the effectiveness of the controllers, about which the graph is also given in Figure 3.1. The reference trajectory is given by $x_r(t) = t$, $y_r(t) = 10\sin(0.1t)$. The demanding distances corresponding to each robot are $\rho_{1x} = 3$, $\rho_{2x} = 3$, $\rho_{3x} = 6$, $\rho_{4x} = 6$, $\rho_{1y} = 0$, $\rho_{2y} = 3$, $\rho_{3y} = 0$, $\rho_{4y} = 3$. The parameters of the robots under simulation are as follows. $b_i = 0.75$, $d_i = 0.3$, $r_i = 0.25$, $m_{ci} = 10$, $m_{wi} = 1$, $I_{ci} = 5.6$, $I_{wi} = 0.005$, $I_{mi} = 0.0025$, $d_{i1} = d_{i2} = 5$. The control parameters are chosen as: $\varepsilon_{i1} = 0.1$, $\varepsilon_{i2} = 0.1$, $k_1 = 2$, $k_2 = 2$, $\gamma_{\theta_{i1}} = \gamma_{\theta_{i2}} = 5$, $\gamma_{r_i} = 4$, $K_i = 2I$, $\Gamma_i = 4I$, parameter ϵ for projection is chosen as $\epsilon = 0.1$. θ_{i1} and θ_{i2} are assumed to be in $[0.15, 0.4]$ and $[0.1, 0.3]$, respectively. The initial values are chosen as: $\hat{\theta}_{i1}(0) = 0.16$, $\hat{\theta}_{i2}(0) = 0.12$, $\hat{\vartheta}_i(0) = [0.05, 3, 3, 0.1, 0, 0.1, 0.01, 0.1, 0.01]^T$, $\hat{w}_{rx,i}(0) = 0.8$, $\hat{c}_{rx} = 0.5$, $\hat{w}_{ry,i}(0) = 1.2$, and $\hat{c}_{ry}(0) = 0.5$. The positions of the four robots and their respective orientation tracking errors are shown in Figures 10.2 and 10.3. Clearly, these results are consistent with those stated in Theorem 10.1 and therefore illustrate our theoretical findings.

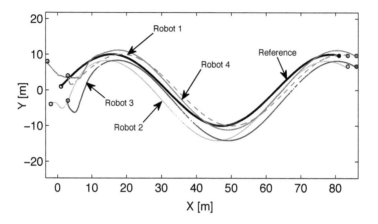

FIGURE 10.2   The positions of the 4 mobile robots in X-Y plane.

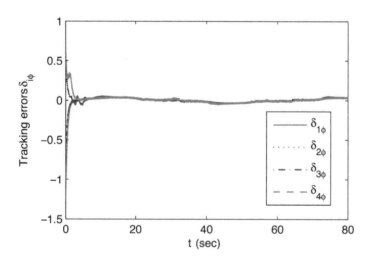

FIGURE 10.3   The tracking errors corresponding to the orientations of the 4 mobile robots.

10.5 NOTES

In this chapter, we have applied the distributed control strategy in Chapter 3 to successfully solve the formation control problem for multiple uncertain nonhonolomic mobile robots with both kinematic model and dynamic model. The control protocol is proposed in this chapter for this challenging problem and the simulation is conducted to validate the effectiveness of the control scheme.

11 Distributed Event-Triggered Adaptive Attitude Synchronization of Multiple Spacecraft

In Chapter 10, we have discussed the distributed adaptive formation control problem for nonholonomic mobile robots. In this chapter, we discuss another class of application example, i.e., spacecraft formation flying. Spacecraft formation flying has gained considerable attention in the past decades, due to its wide applications such as interferometry, earth monitoring, and stellar observation [76]. Attitude synchronization is a fundamental research issue in this area, which aims at reaching a common attitude for a group of spacecraft. A plenty of results on this topic have been reported; see [1, 57, 106] for instance.

In this chapter, we investigate the distributed adaptive attitude synchronization problem for multiple rigid spacecraft with unknown inertial matrices and event-triggered communication under a directed graph condition. The attitude of rigid spacecraft is represented by the Modified Rodriguez Parameters (MRPs). Based on a strongly connected directed graph, two distributed event-triggered adaptive attitude synchronization control schemes are proposed via state/output feedback control. Specifically, to achieve attitude synchronization, distributed reference systems, whose states can achieve consensus based on the designed event triggering communication strategies, are elaborately designed and introduced in the group of spacecraft. Then, a distributed adaptive control law is designed to track the state of the reference system.

11.1 PROBLEM FORMULATION

11.1.1 SPACECRAFT ATTITUDE DYNAMICS

In this chapter, we consider a group of N rigid spacecrafts as shown in Figure11.1.

MRPs are used to describe the attitude of the spacecraft with respect to the inertial frame. Let $\sigma_i = a_i \tan(\frac{\phi_i}{4}) \in \Re^3$ be the MRPs for the ith spacecraft, where a_i and ϕ_i denote the Euler axis and Euler angle, respectively. The attitude kinematics and dynamics of the ith spacecraft is modeled as in [81, 106].

$$\dot{\sigma}_i = F(\sigma_i)\omega_i \tag{11.1}$$

$$J_i\dot{\omega}_i = -\omega_i^\times J_i\omega_i + \tau_i, \ i = 1,\dots,N, \tag{11.2}$$

DOI: 10.1201/9781003394372-11

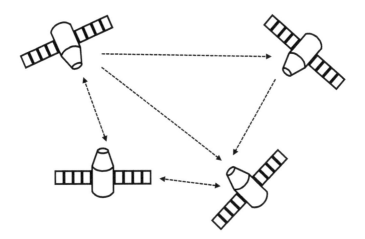

FIGURE 11.1   a group of N rigid spacecrafts.

where $\omega_i \in \Re^3$ is the angular velocity of the ith spacecraft with respect to the inertial frame and expressed in a body-fixed frame. $J_i \in \Re^{3 \times 3}$ denotes the unknown inertia matrix, which is symmetric positive definite. τ_i denotes the control torque. $F(\sigma_i) = \frac{1}{2}\left(\frac{1-\sigma_i^T \sigma_i}{2}I_3 + \sigma_i^\times + \sigma_i \sigma_i^T\right)$, where I_3 is an identity matrix of size 3×3. Notation $x^\times \in \Re^{3 \times 3}$ represents the cross product operator of a vector $x = [x_1, x_2, x_3]^T$, which is defined as

$$\begin{pmatrix} 0 & -x_3 & x_2 \\ x_3 & 0 & -x_1 \\ -x_2 & x_1 & 0 \end{pmatrix}.$$

As stated in [106] and [142], system (11.1)–(11.2) can be transformed into

$$H_i(\sigma_i)\ddot{\sigma}_i + C_i(\sigma_i, \dot{\sigma}_i)\dot{\sigma}_i = F^{-T}(\sigma_i)\tau_i, \tag{11.3}$$

where

$$H_i(\sigma_i) = F^{-T}(\sigma_i)J_i F^{-1}(\sigma_i),$$
$$C_i(\sigma_i, \dot{\sigma}_i) = -F^{-T}(\sigma_i)J_i F^{-1}(\sigma_i)\dot{F}(\sigma_i)F^{-1}(\sigma_i)$$
$$\quad - F^{-T}(\sigma_i)(J_i F^{-1}(\sigma_i)\dot{\sigma}_i)^\times F^{-1}(\sigma_i).$$

Some basic properties of system (11.3) are summarized as follows [81, 142].

Property 11.1 $H_i(\sigma_i)$ *is a symmetric positive definite matrix, which satisfies*

$$J_{i1}\|x\|^2 \le x^T H_i(\sigma_i)x \le J_{i2}(\|\sigma_i\|)\|x\|^2, \forall x \in \Re^3 \tag{11.4}$$

where J_{i1} is a positive constant and $J_{i2}(\|\sigma_i\|)$ is a positive nondecreasing function.

Property 11.2 $\dot{H}_i(\sigma_i) - 2C_i(\sigma_i, \dot{\sigma}_i)$ *is a skew-symmetric matrix, which satisfies* $x^T[\dot{H}_i(\sigma_i) - 2C_i(\sigma_i, \dot{\sigma}_i)]x = 0$, $\forall x \in \Re^3$.

Property 11.3 *The dynamics of system (11.3) is linearly parameterizable, i.e.,* $H_i(\sigma_i)\dot{x} + C_i(\sigma_i, \dot{\sigma}_i)x = Y(\sigma_i, \dot{\sigma}_i, x, \dot{x})\theta_i$, $\forall x \in \Re^3$, *where* $Y(\sigma_i, \dot{\sigma}_i, x, \dot{x}) \in \Re^{3 \times 6}$ *is termed the regression matrix and* $\theta_i \in \Re^6$ *is a constant vector of unknown parameters.*

Remark 11.1 *Note that when the Euler angle* ϕ_i *approaches* 2π, *the singularity arises, which leads to the case that the attitude* σ_i *goes to infinity. This issue can be addressed, by using the MRP shadow set* $\sigma_i^S = -\frac{\sigma_i}{\sigma_i^T \sigma_i}$ *as stated in [106]. In this chapter, the singularity is avoided by ensuring the boundedness of the attitude* σ_i, *thereby keep the Euler angle* ϕ_i *away from* 2π.

Remark 11.2 *Similar to [106] and [81], we assume that the vectors involved in the designed control laws have been appropriately transformed and represented in the same coordinate frame throughout this chapter. That is, the rotation matrices among neighboring spacecraft are first computed, then they will be used to perform coordinate transformations.*

11.1.2   CONTROL OBJECTIVE

The *control objective* is to design distributed adaptive control laws and triggering conditions for a group of spacecraft by using only locally available information such that all the spacecraft can be synchronized at a common attitude with zero angular velocity, i.e., $\lim_{t \to \infty} [\sigma_i(t) - \sigma_j(t)] = \mathbf{0}_3$, $\lim_{t \to \infty} \omega_i(t) = \mathbf{0}_3$ with $\mathbf{0}_3 = [0, 0, 0]^T$ for $i, j \in \mathcal{V}$.

To achieve the control objective, an assumption regarding the graph $\mathcal{G}$ is imposed.

Assumption 11.1.1 *The communication digraph* $\mathcal{G}$ *is strongly connected.*

11.2   THE CASE WITH AVAILABLE ANGULAR VELOCITY MEASUREMENTS

11.2.1   DESIGN OF DISTRIBUTED REFERENCE SYSTEMS

For spacecraft i, we use $t_0^i, t_1^i, \ldots, t_{k_i}^i, \ldots$ with $0 = t_0^i < t_1^i < \ldots < t_{k_i}^i < \ldots < \infty$ for $k_i = 0, 1, 2 \ldots$ and $i \in \mathcal{V}$ to denote the sequence of time instants at which spacecraft i transmits its signals to its out-neighbors. $t_0^i = 0$ is the initial time instant when a group of spacecraft are started. $t_{k_i}^i$ represents the event time instant.

To achieve the attitude synchronization, the following reference system is introduced in each spacecraft i.

$$\dot{z}_i(t) = -\sum_{j=1}^{N} a_{ij}[\bar{z}_i(t) - \bar{z}_j(t)], \qquad (11.5)$$

where $z_i(t) \in \Re^3$ is the state of system (11.5). $\bar{z}_i(t)$ and $\bar{z}_j(t)$ for $i \in \mathcal{V}$ and $j \in \mathcal{N}_i$ respectively represent the latest triggered signals of z_i and z_j, which can be expressed as $\bar{z}_i(t) = z_i(t^i_{k_i})$ for $t \in [t^i_{k_i}, t^i_{k_i+1})$ and $\bar{z}_j(t) = z_j(t^j_{k_j})$ for $t \in [t^j_{k_j}, t^j_{k_j+1})$. Note that for spacecraft i, the received neighboring signals $z_j(t^j_{k_j})$, $j \in \mathcal{N}_i$, are updated only at the time instants $t^j_{k_j}$ for $k_j = 0, 1, 2 \ldots$. The initial condition of system (11.5) is set as $z_i(0) = \sigma_i(0)$, which aims at achieving the weighted average attitude synchronization.

We define a measurement error as $\epsilon_i(t) = z_i(t) - \bar{z}_i(t)$. The triggering condition for spacecraft i is designed as

$$t^i_{k_i+1} = \inf \left\{ t > t^i_{k_i} \mid \|\epsilon_i(t)\|^2 > \Pi_i \right\}, \tag{11.6}$$

where $\Pi_i = \frac{\pi_i}{24\Delta_i} \sum_{j=1}^N a_{ij} \|\bar{z}_i(t) - \bar{z}_j(t)\|^2 + \frac{1}{3\Delta_i} \varpi_i(t)$ and $\varpi_i(t) = \varsigma_i e^{-\iota_i t}$. Δ_i is the diagonal element of the in-degree matrix Δ. π_i, ς_i, and ι_i are positive constants with $0 < \pi_i < 1$. The term $\varpi_i(t) = \varsigma_i e^{-\iota_i t}$ is introduced into the triggering condition (11.6) such that the Zeno behavior in the sense of Definition 8.1 can be excluded.

Remark 11.3 *It is worth mentioning that in essence, the group of reference systems can be regarded as signal-integrator multi-agent systems. Note that the event-triggered consensus problem for such systems has been widely investigated, as can be observed in [18, 88, 117]. However, there still exist some limitations. For example, i) the graphs are undirected or balanced graph; ii) each agent needs to continuously monitor its neighboring states. Unlike these results, the graph considered in this chapter is relaxed to be strongly connected directed graph. Besides, only the triggered neighboring signals are utilized to design the triggering condition (11.6). Thus continuous monitoring of neighboring states can be avoided.*

Lemma 11.1 *Consider a group of distributed reference systems (11.5) under Assumption 11.1.1. With the designed triggering conditions (11.6), the states z_i, $i \in \mathcal{V}$, are globally uniformly bounded and can be synchronized at a common constant vector. In particular, $\lim_{t \to \infty} z_i(t) = \sum_{i=1}^N \xi_i z_i(0) = \sum_{i=1}^N \xi_i \sigma_i(0)$, where ξ_i is a positive graph parameter defined in Lemma 2.1. Moreover, the Zeno behavior in the sense of Definition 8.1 can be excluded.*

Proof. We define a Lyapunov function as

$$V_z = \frac{1}{2} \sum_{i=1}^N \xi_i z_i^T z_i, \tag{11.7}$$

where ξ_i is a positive constant defined in Lemma 2.1.

With $\epsilon_i = z_i - \bar{z}_i$ and (11.5), the derivative of V_z can be computed as

$$\dot{V}_z = -\sum_{i=1}^N \xi_i (\epsilon_i + \bar{z}_i)^T \sum_{j=1}^N a_{ij} (\bar{z}_i - \bar{z}_j)$$

$$\leq \sum_{i=1}^{N} \xi_i \|\epsilon_i\| \sum_{j=1}^{N} a_{ij} \|\bar{z}_i - \bar{z}_j\| - \sum_{i=1}^{N} \xi_i \bar{z}_i^T \sum_{j=1}^{N} a_{ij} (\bar{z}_i - \bar{z}_j). \tag{11.8}$$

Since the digraph is strongly connected and $\xi^T \mathcal{L} = 0$, the property that $\xi_i \sum_{j=1}^{N} a_{ij} = \sum_{j=1}^{N} \xi_j a_{ji}$ holds [162]. Hence it is straightforward to get

$$\sum_{i=1}^{N} \xi_i \bar{z}_i^T \sum_{j=1}^{N} a_{ij}(\bar{z}_i - \bar{z}_j) = \sum_{i=1}^{N} \sum_{j=1}^{N} \xi_i a_{ij} \bar{z}_j^T (\bar{z}_j - \bar{z}_i). \tag{11.9}$$

Substituting (11.9) into (11.8) yields that

$$\dot{V}_z = \sum_{i=1}^{N} \sum_{j=1}^{N} a_{ij} \xi_i \|\epsilon_i\| \|\bar{z}_i - \bar{z}_j\| - \frac{1}{2} \sum_{i=1}^{N} \sum_{j=1}^{N} \xi_i a_{ij} \|\bar{z}_i - \bar{z}_j\|^2$$

$$\leq \sum_{i=1}^{N} \sum_{j=1}^{N} \xi_i a_{ij} \left(2\|\epsilon_i\|^2 - \frac{1}{8} \|\bar{z}_i - \bar{z}_j\|^2 \right) - \frac{1}{4} \sum_{i=1}^{N} \sum_{j=1}^{N} \xi_i a_{ij} \|\bar{z}_i - \bar{z}_j\|^2, \tag{11.10}$$

where $\|\epsilon_i\| \|\bar{z}_i - \bar{z}_j\| \leq 2\|\epsilon_i\|^2 + \frac{1}{8} \|\bar{z}_i - \bar{z}_j\|^2$ has been used.

The term $-\frac{1}{4} \sum_{i=1}^{N} \sum_{j=1}^{N} \xi_i a_{ij} \|\bar{z}_i - \bar{z}_j\|^2$ in (11.10) can be computed as

$$-\frac{1}{4} \sum_{i=1}^{N} \sum_{j=1}^{N} \xi_i a_{ij} \|\bar{z}_i - \bar{z}_j\|^2$$

$$= -\frac{1}{4} \sum_{i=1}^{N} \sum_{j=1}^{N} \xi_i a_{ij} \|z_i - z_j\|^2 - \frac{1}{4} \sum_{i=1}^{N} \sum_{j=1}^{N} \xi_i a_{ij} \|\epsilon_i - \epsilon_j\|^2$$

$$+ \frac{1}{2} \sum_{i=1}^{N} \sum_{j=1}^{N} \xi_i a_{ij} \|z_i - z_j\| \|\epsilon_i - \epsilon_j\|$$

$$\leq -\frac{1}{8} \sum_{i=1}^{N} \sum_{j=1}^{N} \xi_i a_{ij} \|z_i - z_j\|^2 + \frac{1}{4} \sum_{i=1}^{N} \sum_{j=1}^{N} \xi_i a_{ij} \|\epsilon_i - \epsilon_j\|^2$$

$$\leq -\frac{1}{8} \sum_{i=1}^{N} \sum_{j=1}^{N} \xi_i a_{ij} \|z_i - z_j\|^2 + \frac{1}{2} \sum_{i=1}^{N} \sum_{j=1}^{N} \xi_i a_{ij} (\|\epsilon_i\|^2 + \|\epsilon_j\|^2)$$

$$\leq -\frac{1}{8} \sum_{i=1}^{N} \sum_{j=1}^{N} \xi_i a_{ij} \|z_i - z_j\|^2 + \sum_{i=1}^{N} \sum_{j=1}^{N} \xi_i a_{ij} \|\epsilon_i\|^2, \tag{11.11}$$

where the property that $\xi_i \sum_{j=1}^{N} a_{ij} = \sum_{j=1}^{N} \xi_j a_{ji}$ has been used.

Substituting (11.11) into (11.10), we have

$$\dot{V}_z \leq \sum_{i=1}^{N} \sum_{j=1}^{N} \xi_i a_{ij} \left(3\|\epsilon_i\|^2 - \frac{1}{8} \|\bar{z}_i - \bar{z}_j\|^2 \right)$$

$$-\frac{1}{8}\sum_{i=1}^{N}\sum_{j=1}^{N}\xi_i a_{ij}\|z_i - z_j\|^2. \tag{11.12}$$

Let $z = [z_1^T, \ldots, z_N^T]^T$. From the definition of the triggering condition (11.6), (11.12) can be further derived as

$$\dot{V}_z \leq -\frac{1}{8}\sum_{i=1}^{N}\sum_{j=1}^{N}\xi_i a_{ij}\|z_i - z_j\|^2 + \sum_{i=1}^{N}\xi_i \varpi_i(t)$$

$$\leq -\frac{1}{8}z^T(\hat{\mathcal{L}} \otimes I_3)z + \sum_{i=1}^{N}\xi_i \varpi_i(t), \tag{11.13}$$

where $\otimes$ denotes the standard Kronecker product.

Integrating both sides of (11.13) gets

$$V_z(t) + \frac{1}{8}\int_0^t z^T(s)(\hat{\mathcal{L}} \otimes I_3)z(s)ds \leq V_z(0) + \sum_{i=1}^{N}\xi_i \bar{\varpi}_i, \tag{11.14}$$

where $\bar{\varpi}_i = \int_0^t \varpi_i(s)ds$ is a positive bounded function.

From the definition of V_z and (11.14), we can conclude that z_i, $i \in \mathcal{V}$, are uniformly bounded. From (11.5), $\dot{z}_i$ is bounded, implying that $z^T(\hat{\mathcal{L}} \otimes I_3)\dot{z}$ is bounded. By Barbalat's lemma, we can conclude that $\lim_{t\to\infty} z(t)^T(\hat{\mathcal{L}} \otimes I_3)z(t) = 0$. Since the graph is strongly connected, $\hat{\mathcal{L}}$ is symmetric semi-positive definite as stated in Lemma 2.1. Hence $\lim_{t\to\infty}[z_i(t) - z_j(t)] = 0_3$, $i, j \in \mathcal{V}$, which implies that all the states of the reference system (11.5) can be synchronized at a common vector.

Next, we proceed to derive the final synchronization equilibrium. Let $\bar{z} = [\bar{z}_1^T, \ldots, \bar{z}_N^T]^T$. Then, (11.5) can be rewritten as

$$\dot{z} = (\mathcal{L} \otimes I_3)\bar{z}. \tag{11.15}$$

Multiplying both sides of (11.15) by $(\xi \otimes I_3)^T$, we get

$$(\xi \otimes I_3)^T \dot{z} = (\xi \otimes I_3)^T(\mathcal{L} \otimes I_3)\bar{z} = (\xi^T \mathcal{L} \otimes I_3)\bar{z} = 0_3. \tag{11.16}$$

Integrating both sides of (11.16), we have

$$\lim_{t\to\infty}\sum_{i=1}^{N}\xi_i z_i(t) = \sum_{i=1}^{N}\xi_i z_i(0). \tag{11.17}$$

With $\lim_{t\to\infty}[z_i(t) - z_j(t)] = 0_3$, $\sum_{i=1}^{N}\xi_i = 1$, and $z_i(0) = \sigma_i(0)$, the final synchronization equilibrium can be represented as

$$\lim_{t\to\infty} z_i(t) = \sum_{i=1}^{N}\xi_i z_i(0) = \sum_{i=1}^{N}\xi_i \sigma_i(0). \tag{11.18}$$

In what follows, we show that the Zeno behavior in each spacecraft can be excluded. Assume that the triggering condition (11.6) triggers at time instants $t_{k_i}^i$, $k_i = 0, 1, 2 \ldots$, $i \in \mathcal{V}$. Hence $\epsilon_i(t_{k_i}^i) = 0$. During the time interval $[t_{k_i}^i, t_{k_i+1}^i)$, the derivative of ϵ_i can be computed as $\dot{\epsilon}_i = \dot{z}_i = -\sum_{j=1}^{N} a_{ij}(\bar{z}_i - \bar{z}_j)$. Then, integrating this equation gets $\epsilon_i = \int_{t_{k_i}^i}^{t} \dot{z}_i(s) ds = \int_{t_{k_i}^i}^{t} [-\sum_{j=1}^{N} a_{ij}(\bar{z}_i(s) - \bar{z}_j(s))] ds$. Since $\bar{z}_i$ and $\bar{z}_j$, $i, j \in \mathcal{V}$, are bounded, there exists a positive constant $\bar{\varsigma}_i$ such that $\|\dot{z}_i\| \leq \bar{\varsigma}_i$. Hence, we have $\epsilon_i \leq (t - t_{k_i}^i) \bar{\varsigma}_i$. From the definition of the triggering condition (11.6), it implies that the next event will not be triggered before $\|\epsilon_i\|^2 = \frac{\pi_i}{24\Delta_i} \sum_{j=1}^{N} a_{ij} \|\bar{z}_i - \bar{z}_j\|^2 + \frac{1}{3\Delta_i} \varpi_i(t)$. We define the inter-event time interval as $\tau_{k_i}^i = t_{k_i+1}^i - t_{k_i}^i$. Therefore, the inter-event time interval can be represented as

$$\tau_{k_i}^i = \frac{\sqrt{\frac{\pi_i}{24\Delta_i} \sum_{j=1}^{N} a_{ij} \|\bar{z}_i(t_{k_i}^i) - \bar{z}_j(t_{k_j}^j)\|^2 + \frac{1}{3\Delta_i} \varpi_i(t_{k_i}^i + \tau_{k_i}^i)}}{\bar{\varsigma}_i}. \tag{11.19}$$

Since $\varpi_i(t)$ is an exponentially decaying function and always greater than zero, $\tau_{k_i}^i$ is always greater than zero for any finite time instant $t_{k_i}^i$. Therefore, the Zeno behavior, as stated in Definition 8.1, is ruled out in each spacecraft.

Remark 11.4 *The reference system introduced in each spacecraft plays an essential role in the attitude synchronization of multiple spacecraft. It generates a reference signal for each spacecraft to track. With the aid of the reference system, the attitude synchronization problem for multiple spacecraft can be converted to two sub-problems, namely, the synchronization problem for a group of reference systems and the attitude tracking problem for single spacecraft. If all the states of reference systems can achieve synchronization and each spacecraft can track its reference signal asymptotically, then the attitude synchronization problem for multiple spacecraft is solved. However, it is nontrivial to solve the former problem due to the fact that the Laplacian matrix associated with a strongly connected directed graph is not symmetric positive semi-definite. To solve this issue, a graph-parameter-related Lyapunov function (11.7) is chosen in the proof of Lemma 11.1, based on which a triggering condition (11.6) is elaborately designed.*

11.2.2 DESIGN OF DISTRIBUTED ADAPTIVE CONTROL LAWS

For notational convenience, the time variable t for signals is omitted hereafter if there is no confusion. We define a sliding mode error and a tracking error, respectively, as $s_i = \sigma_i + \dot{\sigma}_i$ and $e_i = s_i - z_i$. With Property 11.3 and (11.5), we can get

$$H_i(\sigma_i)\dot{e}_i + C_i(\sigma_i, \dot{\sigma}_i)e_i$$
$$= H_i(\sigma_i)(\dot{\sigma}_i - \dot{z}_i) + C_i(\sigma_i, \dot{\sigma}_i)(\sigma_i - z_i) + F^{-T}(\sigma_i)\tau_i$$
$$= Y(\sigma_i, \dot{\sigma}_i, \sigma_i - z_i, \dot{\sigma}_i - \dot{z}_i)\theta_i + F^{-T}(\sigma_i)\tau_i. \tag{11.20}$$

The actual control law τ_i and parameter update law for $\hat{\theta}_i$ are respectively designed as

$$\tau_i = F^T(\sigma_i)\bar{\tau}_i, \tag{11.21}$$

$$\bar{\tau}_i = -Y(\sigma_i, \dot{\sigma}_i, \sigma_i - z_i, \dot{\sigma}_i - \dot{z}_i)\hat{\theta}_i - K_i e_i, \tag{11.22}$$

$$\dot{\hat{\theta}}_i = \Gamma_i Y(\sigma_i, \dot{\sigma}_i, \sigma_i - z_i, \dot{\sigma}_i - \dot{z}_i)^T e_i, \tag{11.23}$$

where $\hat{\theta}_i$ is the estimate of the unknown parameters θ_i. $K_i \in \Re^{3\times3}$ and $\Gamma_i \in \Re^{6\times6}$ are symmetric positive definite matrices.

Remark 11.5 *From (11.21)–(11.23), it can be observed that only the local continuous signals and neighboring triggered signals $\bar{z}_j$ for $j \in \mathcal{N}_i$, contained in $\dot{z}_i = -\sum_{j=1}^{N} a_{ij}(\bar{z}_i - \bar{z}_j)$, are utilized to design adaptive attitude synchronization control law and parameter update law. Therefore, the developed attitude synchronization algorithm is fully distributed.*

11.2.3 STABILITY AND SYNCHRONIZATION ANALYSIS

The main results in this section are formally stated in the following theorem.

Theorem 11.1 *Consider a group of N rigid spacecraft (11.1)–(11.2) under Assumption 11.1.1. With the designed triggering conditions (11.6), reference systems (11.5), distributed adaptive control laws (11.21)–(11.22) and parameter update laws (11.23), all the closed-loop signals are uniformly bounded and all the spacecraft can be synchronized at a common attitude with zero angular velocity. In particular, $\lim_{t\to\infty} \sigma_i(t) = \sum_{i=1}^{N} \xi_i \sigma_i(0)$ and $\lim_{t\to\infty} \omega_i(t) = \mathbf{0}_3$.*

Proof. By Lemma 11.1 and (11.5), z_i and $\dot{z}_i$ are bounded and the states of the reference systems can be synchronized at a common constant vector, i.e., $\lim_{t\to\infty} z_i(t) = \sum_{i=1}^{N} \xi_i \sigma_i(0)$. In what follows, we will show that the attitude of each spacecraft can track the state of its reference system asymptotically, i.e., $\lim_{t\to\infty} \sigma_i(t) = \lim_{t\to\infty} z_i(t) = \sum_{i=1}^{N} \xi_i \sigma_i(0)$.

Define a Lyapunov function as

$$V = \frac{1}{2}\sum_{i=1}^{N} e_i^T H_i(\sigma_i) e_i + \frac{1}{2}\sum_{i=1}^{N} \tilde{\theta}_i^T \Gamma_i^{-1} \tilde{\theta}_i, \tag{11.24}$$

where $\tilde{\theta}_i = \theta_i - \hat{\theta}_i$ denotes an estimation error for θ_i.

From (11.21)–(11.23) and Property 11.2, the derivative of V can be computed as

$$\dot{V} = -\sum_{i=1}^{N} e_i^T K_i e_i + \sum_{i=1}^{N} \tilde{\theta}_i^T \Gamma_i^{-1}\left(\Gamma_i \bar{Y}^T e_i - \dot{\hat{\theta}}_i\right)$$

$$= -\sum_{i=1}^{N} e_i^T K_i e_i, \tag{11.25}$$

where for notational convenience, $\bar{Y} = Y(\sigma_i, \dot{\sigma}_i, \sigma_i - z_i, \dot{\sigma}_i - \dot{z}_i)$ has been used.

From the definition of V and (11.25), it can be established that e_i and $\hat{\theta}_i$ are bounded for spacecraft i. With $e_i = s_i - z_i$, s_i is bounded. In view of $s_i = \sigma_i + \dot{\sigma}_i$, σ_i and $\dot{\sigma}_i$ are bounded. Thus the regressor matrix $Y(\sigma_i, \dot{\sigma}_i, \sigma_i - z_i, \dot{\sigma}_i - \dot{z}_i)$ is bounded. From (11.21) and (11.22), it implies that the control law of each spacecraft is bounded. Therefore, we can conclude that all the closed-loop signals are uniformly bounded.

We now set out to analyze the final synchronization attitude. By Lemma 11.1, we have $\lim_{t\to\infty} z_i(t) = \sum_{i=1}^{N} \xi_i \sigma_i(0)$. By LaSalle-Yoshizawa theorem, it follows from (11.25) that $\lim_{t\to\infty} e_i(t) = 0_3$, which implies that s_i can track the reference signal z_i asymptotically. As a result, we can obtain that $\lim_{t\to\infty} s_i(t) = \lim_{t\to\infty} z_i(t) = \sum_{i=1}^{N} \xi_i \sigma_i(0)$. Recalling the definition of the sliding mode variable $s_i = \sigma_i + \dot{\sigma}_i$, it can also be seen as an ISS linear system $\dot{\sigma}_i = -\sigma_i + s_i$ with s_i being its input. Due to the fact that a constant vector s_i leads to a constant vector σ_i and $\lim_{t\to\infty} \sigma_i(t) = \lim_{t\to\infty} s_i(t)$, the final synchronization attitude σ_i can be represented as $\lim_{t\to\infty} \sigma_i(t) = \sum_{i=1}^{N} \xi_i \sigma_i(0)$. Furthermore, it follows from (11.1) that $\lim_{t\to\infty} \omega_i(t) = 0_3$.

Remark 11.6 *In the attitude synchronization problem for multiple spacecraft without a leader, a concerned issue is to confirm the final attitude that the group of spacecraft can be eventually synchronized to. As shown in Lemma 11.1 and Theorem 11.1, the final attitude in this chapter is closely dependent on the initial conditions of the reference systems, which can be represented by $\lim_{t\to\infty} \sigma_i(t) = \sum_{i=1}^{N} \xi_i z_i(0)$. This implies that if the graph among spacecraft is given and in turn ξ_i can be computed, we can determine the final synchronization attitude by properly choosing the initial conditions of reference systems. As a special case, in this chapter they are chosen to be equal to the initial attitudes of all spacecraft. Consequently, the weighted average attitude synchronization is achieved, i.e., $\lim_{t\to\infty} \sigma_i(t) = \sum_{i=1}^{N} \xi_i \sigma_i(0)$.*

11.3   THE CASE WITHOUT ANGULAR VELOCITY MEASUREMENTS

In previous section, fully distributed adaptive attitude synchronization algorithms are developed for a group of rigid spacecraft. To generate distributed adaptive control laws, the angular velocity is used, which requires each spacecraft to be equipped with an angular velocity sensor. Undoubtedly, such attitude synchronization algorithms will increase the implementation cost. In what follows, we shall design a distributed adaptive output feedback synchronization algorithm.

11.3.1 MODIFICATION OF DISTRIBUTED REFERENCE SYSTEMS

In this section, the reference system introduced in each spacecraft i is modified as follows.

$$\dot{z}_{i,1}(t) = -c_{i,1}[z_{i,1}(t) - z_{i,2}(t)], \tag{11.26}$$

$$\dot{z}_{i,2}(t) = -c_{i,2}[z_{i,2}(t) - z_{i,3}(t)], \tag{11.27}$$

$$\dot{z}_{i,3}(t) = -\sum_{j=1}^{N} a_{ij}[\bar{z}_{i,3}(t) - \bar{z}_{j,3}(t)], \tag{11.28}$$

$$y_{i,r}(t) = z_{i,1}(t), \tag{11.29}$$

where $z_{i,q}(t) \in \Re^3$, $q = 1, 2, 3$, and $y_{i,r}(t) \in \Re^3$, $i \in \mathcal{V}$, are the states and output of the reference system (11.26)–(11.29), respectively. Note that the subscript i, r in $y_{i,r}(t)$ represents the reference signal generated for spacecraft i. $c_{i,1}$ and $c_{i,2}$ are positive constants. $\bar{z}_{i,3}(t) = z_{i,3}(t_{k_i}^i)$ for $t \in [t_{k_i}^i, t_{k_i+1}^i)$ and $\bar{z}_{j,3}(t) = z_{j,3}(t_{k_j}^j)$ for $t \in [t_{k_j}^j, t_{k_j+1}^j)$. The initial conditions of reference system (11.26)–(11.29) are chosen as $z_{i,1}(0) = z_{i,2}(0) = z_{i,3}(0) = \sigma_i(0)$ to achieve weighted average attitude synchronization.

Remark 11.7 *Different from the reference system in (11.5), the reference system in this section has been modified to be a third-order cascade system (11.26)–(11.29). By doing this, a reference signal, whose first third-order derivatives (i.e., $\dot{y}_{i,r}$, $\ddot{y}_{i,r}$, and $\dddot{y}_{i,r}$) exist and are bounded, can be generated. This is a precondition for the implementation of the designed distributed adaptive output feedback control law in the sequel. More discussions will be provided in Remark 11.8.*

Define a measurement error as $\epsilon_{i,3}(t) = z_{i,3}(t) - \bar{z}_{i,3}(t)$. The triggering condition for spacecraft i is designed as

$$t_{k_i+1}^i = \inf\left\{t > t_{k_i}^i \,\middle|\, \|\epsilon_{i,3}(t)\|^2 > \Pi_{i,3}\right\}, \tag{11.30}$$

where $\Pi_{i,3} = \frac{\pi_i}{24\Delta_i} \sum_{j=1}^{N} a_{ij}\|\bar{z}_{i,3}(t) - \bar{z}_{j,3}(t)\|^2 + \frac{1}{3\Delta_i}\varpi_i(t)$ and $\varpi_i(t) = \varsigma_i e^{-\iota_i t}$. Δ_i is the diagonal element of in-degree matrix Δ. π_i, ς_i, and ι_i are positive constants with $0 < \pi_i < 1$.

Lemma 11.2 *Consider a group of distributed reference systems (11.26)–(11.29) with Assumption 11.1.1. Under the designed triggering conditions (11.30), all the states of reference systems are globally uniformly bounded and the output $y_{i,r}$ can be synchronized at a common constant vector. In particular, $\lim_{t\to\infty} y_{i,r}(t) = \sum_{i=1}^{N} \xi_i \sigma_i(0)$. Meanwhile, the Zeno behavior in the sense of Definition 8.1 can be ruled out.*

Proof. From (11.26)–(11.29), it can be observed that the reference system is a cascade system consisting of three subsystems and the subsystem (11.28) takes the same form as the reference system (11.5). By Lemma (11.1), we can obtain that $z_{i,3}$ and $\dot{z}_{i,3}$ are globally uniformly bounded, $\lim_{t\to\infty} z_{i,3}(t) = \sum_{i=1}^{N} \xi_i z_{i,3}(0) = \sum_{i=1}^{N} \xi_i \sigma_i(0)$ and the Zeno behavior in the sense of Definition 8.1 can be excluded. Hence the remaining task is to show that the state $z_{i,1}$ can track the state $z_{i,3}$ asymptotically. To this end, we first show that $z_{i,2}$ can track $z_{i,3}$ asymptotically and then show that $z_{i,1}$ can track $z_{i,2}$ asymptotically.

We define a tracking error as $\delta_{i,2} = z_{i,2} - z_{i,3}$. Choose a Lyapunov function as $V_{\delta_2} = \frac{1}{2}\delta_{i,2}^T \delta_{i,2}$, whose derivative can be computed as

$$\dot{V}_{\delta_2} = -c_{i,2}\delta_{i,2}^T \delta_{i,2} - \delta_{i,2}^T \dot{z}_{i,3}$$

$$= -c_{i,2}\|\delta_{i,2}\|^2 + \|\delta_{i,2}\|\|\sum_{j=1}^{N} a_{ij}(\bar{z}_{i,3} - \bar{z}_{j,3})\|$$

$$\leq -\frac{c_{i,2}}{2}\|\delta_{i,2}\|^2 + \frac{1}{2c_{i,2}}\|\sum_{j=1}^{N} a_{ij}(\bar{z}_{i,3} - \bar{z}_{j,3})\|^2$$

$$\leq -\frac{c_{i,2}}{2}\|\delta_{i,2}\|^2 + \frac{N}{2c_{i,2}}\sum_{j=1}^{N} a_{ij}\|(\bar{z}_{i,3} - \bar{z}_{j,3})\|^2, \qquad (11.31)$$

where the Cauchy inequality $\|a_1 + a_2 + \ldots + a_N\|^2 \leq N(\|a_1\|^2 + \|a_2\|^2 + \ldots + \|a_N\|^2)$, where a_i, $i = 1, \ldots, N$, are constant vectors with appropriate dimension, has been used.

By integrating both sides of (11.31), we have

$$V_{\delta_2}(t) + \frac{c_{i,2}}{2}\int_0^t \|\delta_{i,2}(s)\|^2 ds$$

$$\leq V_{\delta_2}(0) + \frac{N}{2c_{i,2}}\int_0^t \sum_{j=1}^{N} a_{ij}\|(\bar{z}_{i,3}(s) - \bar{z}_{j,3}(s))\|^2 ds. \qquad (11.32)$$

From (11.14), it follows that $\int_0^t \sum_{j=1}^{N} a_{ij}\|(\bar{z}_{i,3}(s) - \bar{z}_{j,3}(s))\|^2 ds$ is a bounded function. From the definition of V_{δ_2} and (11.32), we can conclude that $\delta_{i,2}$ and $\int_0^t \|\delta_{i,2}(s)\|^2 ds$ are bounded. In view of $\delta_{i,2} = z_{i,2} - z_{i,3}$, $z_{i,2}$ is bounded. Furthermore, we can compute that $\dot{\delta}_{i,2} = -c_{i,2}\delta_{i,2} - \dot{z}_{i,3}$, which implies that $\dot{\delta}_{i,2}$ is bounded. By Barbalat' Lemma, it follows that $\lim_{t\to\infty} \delta_{i,2}(t) = 0_3$. That is, $z_{i,2}$ can track $z_{i,3}$ asymptotically, i.e., $\lim_{t\to\infty} z_{i,2}(t) = \lim_{t\to\infty} z_{i,3}(t) = \sum_{i=1}^{N} \xi_i \sigma_i(0)$.

We next show that $z_{i,1}$ can track $z_{i,2}$ asymptotically. Define a tracking error $\delta_{i,1} = z_{i,1} - z_{i,2}$ and a Lyapunov function as $V_{\delta_1} = \frac{1}{2}\delta_{i,1}^T \delta_{i,1}$. From (11.26), the derivative of V_{δ_1} can be derived as

$$\dot{V}_{\delta_1} \leq -\frac{c_{i,1}}{2}\|\delta_{i,1}\|^2 + \frac{1}{2c_{i,1}}\|\delta_{i,2}\|^2. \qquad (11.33)$$

Integrating both sides of (11.33) and following the similar analysis as before, we can get that $\delta_{i,1}$ and $\int_0^t \|\delta_{i,1}(s)\|^2 ds$ are bounded. In view of $\delta_{i,1} = z_{i,1} - z_{i,2}$, $z_{i,1}$ is bounded. Therefore, all the closed-loop signals of the reference system are globally uniformly bounded.

Furthermore, we can get that $\dot{\delta}_{i,1} = -c_{i,1}\delta_{i,1} + c_{i,2}\delta_{i,2}$, implying that $\dot{\delta}_{i,1}$ is bounded. By applying Barbalat' Lemma again, we can get that $\lim_{t\to\infty} \delta_{i,1}(t) = \mathbf{0}_3$. That is, the state $z_{i,1}$ can track the state $z_{i,2}$ asymptotically, i.e., $\lim_{t\to\infty} z_{i,1}(t) = \lim_{t\to\infty} z_{i,2}(t) = \sum_{i=1}^{N} \xi_i\sigma_i(0)$. Hence it follows from (11.29) that $\lim_{t\to\infty} y_{i,r}(t) = \sum_{i=1}^{N} \xi_i\sigma_i(0)$.

11.3.2 DESIGN OF DISTRIBUTED ADAPTIVE OUTPUT FEEDBACK CONTROL LAWS

Since the angular velocity of spacecraft is unmeasurable, an adaptive filter-based output feedback tracking control scheme is adopted in this section [102, 142]. We define a tracking error as $e_i = \sigma_i - y_{i,r}$. For spacecraft i, the following pseudo-velocity filter is introduced.

$$\dot{q}_i = -(k_i + 1)q_i + (k_i^2 + 1)e_i, \qquad (11.34)$$

$$e_{if} = -k_i e_i + q_i, \qquad (11.35)$$

where $q_i \in \Re^3$ and $e_{if} \in \Re^3$ are the state and output of the pseudo-velocity filter, respectively. $k_i = 1 + k_{is}$, where k_{is} is a positive constant to be designed. The initial condition of q_i is set as $q_i(0) = k_i e_i(0)$. Note that e_i, q_i, and e_{if} are all available variables, which can be used to generate the distributed adaptive output feedback control law.

From (11.34) and (11.35), the dynamics of e_{if} can be computed as

$$\dot{e}_{if} = -k_i\eta_i - e_{if} + e_i, \qquad (11.36)$$

where $\eta_i = e_i + \dot{e}_i + e_{if}$ is an auxiliary error variable, which contains the information of angular velocity of spacecraft i.

In view of $e_i = \sigma_i - y_{i,r}$ and (11.36), the derivative of η_i can be computed as

$$\dot{\eta}_i = \dot{e}_i + \ddot{\sigma}_i - \ddot{y}_{i,r} - k_i\eta_i - e_{if} + e_i. \qquad (11.37)$$

Multiplying both sides of (11.37) by $H_i(\sigma_i)$ and substituting (11.3) into it, we have

$$\begin{aligned}
H_i(\sigma_i)\dot{\eta}_i &= H_i(\sigma_i)\dot{e}_i - C_i(\sigma_i, \dot{\sigma}_i)\dot{\sigma}_i - H_i(\sigma_i)\ddot{y}_{i,r} \\
&\quad - k_i H_i(\sigma_i)\eta_i - H_i(\sigma_i)e_{if} + H_i(\sigma_i)e_i \\
&\quad + F^{-T}(\sigma_i)\tau_i \\
&= H_i(\sigma_i)\dot{e}_i - C_i(\sigma_i, \dot{\sigma}_i)(\dot{e}_i + \dot{y}_{i,r}) - H_i(\sigma_i)\ddot{y}_{i,r} \\
&\quad - H_i(\sigma_i)\eta_i - k_{is}H_i(\sigma_i)\eta_i - H_i(\sigma_i)e_{if} \\
&\quad + H_i(\sigma_i)e_i + F^{-T}(\sigma_i)\tau_i \\
&= -C_i(\sigma_i, \dot{\sigma}_i)(\dot{e}_i + \dot{y}_{i,r}) - H_i(\sigma_i)\ddot{y}_{i,r} \\
&\quad - k_{is}H_i(\sigma_i)\eta_i - 2H_i(\sigma_i)e_{if} + F^{-T}(\sigma_i)\tau_i, \qquad (11.38)
\end{aligned}$$

where $\dot{\sigma}_i = \dot{e}_i + \dot{y}_{i,r}$ and $\eta_i = e_i + \dot{e}_i + e_{if}$ have been used.

Based on Property 11.3, we define a desired linear parametrization equation as

$$Y_d(t)\theta_i = H_i(y_{i,r})\ddot{y}_{i,r} + C_i(y_{i,r}, \dot{y}_{i,r})\dot{y}_{i,r}. \tag{11.39}$$

Adding and subtracting (11.39) to the right-hand side of (11.38), we get

$$\begin{aligned}
H_i(\sigma_i)\dot{\eta}_i = {} & C_i(\sigma_i, \dot{\sigma}_i)(e_i + e_{if} - \eta_i) - C_i(\sigma_i, \dot{\sigma}_i)\dot{y}_{i,r} \\
& - H_i(\sigma_i)\ddot{y}_{i,r} + Y_d(t)\theta_i - k_{is}H_i(\sigma_i)\eta_i \\
& - 2H_i(\sigma_i)e_{if} + F^{-T}(\sigma_i)\tau_i - H_i(y_{i,r})\ddot{y}_{i,r} - C_i(y_{i,r}, \dot{y}_{i,r})\dot{y}_{i,r} \\
= {} & - C_i(\sigma_i, \dot{\sigma}_i)\eta_i - k_{is}H_i(\sigma_i)\eta_i - Y_d(t)\theta_i + \chi_i + F^{-T}(\sigma_i)\tau_i, \tag{11.40}
\end{aligned}$$

where

$$\begin{aligned}
\chi_i = {} & C_i(\sigma_i, \dot{\sigma}_i)(e_i + e_{if}) - C_i(\sigma_i, \dot{\sigma}_i)\dot{y}_{i,r} - H_i(\sigma_i)\ddot{y}_{i,r} \\
& - 2H_i(\sigma_i)e_{if} + H_i(y_{i,r})\ddot{y}_{i,r} + C_i(y_{i,r}, \dot{y}_{i,r})\dot{y}_{i,r}. \tag{11.41}
\end{aligned}$$

By Lemma 11.2, we know that all the states of the reference system (11.26)–(11.29) are uniformly bounded. Thus there exist positive constants $\bar{y}_{i1}$, $\bar{y}_{i2}$, and $\bar{y}_{i3}$ such that $\|y_{i,r}\| \leq \bar{y}_{i1}$, $\|\dot{y}_{i,r}\| \leq \bar{y}_{i2}$ and $\|\ddot{y}_{i,r}\| \leq \bar{y}_{i3}$. As shown in [102, 142], the upper bound of χ_i can be represented as

$$\|\chi_i\| \leq \rho_i(\bar{y}_{i1}, \bar{y}_{i2}, \bar{y}_{i3}, \|\zeta_i\|)\|\zeta_i\|, \tag{11.42}$$

where $\zeta_i = [e_i^T, e_{if}^T, \eta_i^T]^T$, $\rho_i(\cdot)$ is a positive nondecreasing implicit function. Note that the choice of $\rho_i(\cdot)$ is non-unique.

The distributed adaptive output feedback control law τ_i and parameter update law for $\hat{\theta}_i$ are designed as

$$\tau_i = F^T(\sigma_i)\bar{\tau}_i, \tag{11.43}$$

$$\bar{\tau}_i = Y_d(t)\hat{\theta}_i + k_i e_{if} - e_i, \tag{11.44}$$

$$\hat{\theta}_i = -\Gamma_i \int_0^t Y_d(s)^T[e_i(s) + e_{if}(s)]ds - \Gamma_i Y_d(t)^T e_i,$$

$$+ \Gamma_i \int_0^t \dot{Y}_d(s)^T e_i(s)ds, \tag{11.45}$$

where $\Gamma_i \in \Re^{6 \times 6}$ is a positive definite matrix. We denote the estimation error of θ_i by $\tilde{\theta}_i = \theta_i - \hat{\theta}_i$. To facilitate the subsequent stability analysis, an equivalent form of (11.45) is given as follows

$$\dot{\hat{\theta}}_i = -\Gamma_i Y_d(t)^T \eta_i. \tag{11.46}$$

Due to the fact that the auxiliary error variable η_i contains the angular velocity, the parameter update law (11.46) is not implementable.

Remark 11.8 *As observed in (11.45), the first third-order derivatives of the reference signal $y_{i,r}$ (i.e., $\dot{y}_{i,r}$, $\ddot{y}_{i,r}$, and $\dddot{y}_{i,r}$), involved in $\dot{Y}_d(t)$, are used in the implementation of parameter update law for $\hat{\theta}_i$. This is the main reason why we need to modify the reference system. By doing this, it can generate a reference signal that is suitable for the subsequent adaptive output feedback control law.*

11.3.3 STABILITY AND SYNCHRONIZATION ANALYSIS

The main results in this section are formally stated in the following theorem.

Theorem 11.2 *Consider a group of N rigid spacecraft (11.1)–(11.2) consisting of the designed triggering conditions (11.30), reference systems (11.26)–(11.29), pseudo-velocity filters (11.35)–(11.34), distributed adaptive output feedback control laws (11.43) and parameter update laws (11.45), under Assumption 11.1.1. If the selected design parameter k_{is} satisfies*

$$k_{is} = \frac{1}{J_{i1}}(1 + k_{in}) \tag{11.47}$$

$$k_{in} \geq \frac{1}{2}\rho_i^2\left(\bar{y}_{i1}, \bar{y}_{i2}, \bar{y}_{i3}, \sqrt{\frac{\lambda_{i2}(\sigma_i(0))}{\lambda_{i1}}}\|v_i(0)\|\right) \tag{11.48}$$

with

$$v_i = [e_i^T, e_{if}^T, \eta_i^T, \tilde{\theta}_i^T]^T \tag{11.49}$$

$$\lambda_{i1} = \frac{1}{2}\min\{1, J_{i1}, \lambda_{\min}(\Gamma_i^{-1})\} \tag{11.50}$$

$$\lambda_{i2}(\sigma_i) = \frac{1}{2}\max\{1, J_{i2}(\sigma_i), \lambda_{\max}(\Gamma_i^{-1})\}, \tag{11.51}$$

then all the closed-loop signals are uniformly bounded and all the spacecraft can be synchronized at a common attitude with zero angular velocity. In particular, $\lim_{t\to\infty}\sigma_i(t) = \sum_{i=1}^{N}\xi_i\sigma_i(0)$ and $\lim_{t\to\infty}\omega_i(t) = \mathbf{0}_3$. Meanwhile, the Zeno behavior in the sense of Definition 8.1 can be ruled out.

Proof. By Lemma 11.2, we know that $z_{i,1}$, $z_{i,2}$, $z_{i,3}$, and $y_{i,r}$ are bounded signals, the Zeno behavior in the sense of Definition 8.1 is excluded and $\lim_{t\to\infty}y_{i,r}(t) = \sum_{i=1}^{N}\xi_i\sigma_i(0)$. It remains to show that the attitude of spacecraft i can track the reference signal $y_{i,r}$ asymptotically.

We define a Lyapnov function as

$$V_t = \frac{1}{2}e_i^T e_i + \frac{1}{2}e_{if}^T e_{if} + \frac{1}{2}\eta_i^T H_i(\sigma_i)\eta_i + \frac{1}{2}\tilde{\theta}_i^T\Gamma_i^{-1}\tilde{\theta}_i, \tag{11.52}$$

where $\tilde{\theta}_i = \theta_i - \hat{\theta}_i$ denotes the estimation error of θ_i. Obviously, the Lyapnov function V_t owns the following property.

$$\lambda_{i1}\|\zeta_i\|^2 \leq \lambda_{i1}\|v_i\|^2 \leq V_t \leq \lambda_{i2}(\sigma_i)\|v_i\|^2, \tag{11.53}$$

where v_i is defined in (11.49).

Using (11.36), (11.40), (11.43), (11.44), (11.46), and Property 11.2, the derivative of V_t can be computed as

$$\dot{V}_t = -e_i^T e_i - e_{if}^T e_{if} - k_{is}\eta_i^T H_i(\sigma_i)\eta_i + \eta_i^T\chi_i$$

$$\leq - \|e_i\|^2 - \|e_{if}\|^2 - k_{is} J_{i1} \|\eta_i\|^2 + \|\eta_i\| \|\chi_i\|. \tag{11.54}$$

Substituting (11.42), (11.47), and (11.48) into (11.54), we have

$$\dot{V}_t \leq - \|e_i\|^2 - \|e_{if}\|^2 - \|\eta_i\|^2 - k_{in} \|\eta_i\|^2,$$
$$+ \|\eta_i\| \rho_i(\bar{y}_{i1}, \bar{y}_{i2}, \bar{y}_{i3}, \|\zeta_i\|) \|\zeta_i\|,$$
$$\leq - \|\zeta_i\|^2 + \frac{\rho_i^2(\bar{y}_{i1}, \bar{y}_{i2}, \bar{y}_{i3}, \|\zeta_i\|)}{4k_{in}} \|\zeta_i\|^2, \tag{11.55}$$

where the Young's inequality $\|a\| \|b\| \leq k\|a\|^2 + \frac{1}{4k}\|b\|^2$, where a and b are vectors with appropriate dimension and k is a positive constant, has been used.

Due to the fact that $\rho_i(\cdot)$ is a positive nondecreasing implicit function, we can further get

$$\dot{V}_t \leq - \left(1 - \frac{\rho_i^2(\bar{y}_{i1}, \bar{y}_{i2}, \bar{y}_{i3}, \sqrt{V_t/\lambda_{i1}})}{4k_{in}} \right) \|\zeta_i\|^2. \tag{11.56}$$

If $k_{in} > \frac{1}{2}\rho_i^2(\bar{y}_{i1}, \bar{y}_{i2}, \bar{y}_{i3}, \sqrt{V_t/\lambda_{i1}})$, then

$$\dot{V}_t \leq -\frac{1}{2}\|\zeta_i\|^2, t \geq 0, \tag{11.57}$$

which implies that $V_t(t) \leq V_t(0)$. Furthermore, if $k_{in} > \frac{1}{2}\rho_i^2(\bar{y}_{i1}, \bar{y}_{i2}, \bar{y}_{i3}, \sqrt{V_t(0)/\lambda_{i1}})$, then the inequality (11.57) can also be ensured. Hence it follows from (11.53) that a sufficient condition to guarantee the inequality (11.57) can be given as

$$k_{in} \geq \frac{1}{2}\rho_i^2 \left(\bar{y}_{i1}, \bar{y}_{i2}, \bar{y}_{i3}, \sqrt{\frac{\lambda_{i2}(\sigma_i(0))}{\lambda_{i1}}} \|v_i(0)\| \right). \tag{11.58}$$

From the definition of V_t and (11.57), it can be concluded that e_i, e_{if}, η_i, and $\hat{\theta}_i$ are uniformly bounded. From (11.34), it follows that q_i is bounded. Since $y_{i,r}$ is bounded, the boundedness of σ_i can be ensured. From (11.39), $Y_d(t)$ is bounded. It follows from (11.43) and (11.44) that τ_i is bounded. Therefore, all the closed-loop signals are uniformly bounded. By LaSalle-Yoshizawa theorem, it follows from (11.57) that $\lim_{t \to \infty} \zeta_i(t) = 0$. As a result, $\lim_{t \to \infty} e_i(t) = 0$, which means that the attitude σ_i can track the reference signal $y_{i,r}$ asymptotically. Hence $\lim_{t \to \infty} \sigma_i(t) = \lim_{t \to \infty} y_{i,r}(t) = \sum_{i=1}^{N} \xi_i \sigma_i(0)$. Furthermore, it follows from (11.1) that $\lim_{t \to \infty} \omega_i(t) = 0_3$. Therefore, we can conclude that all the closed-loop signals are uniformly bounded and all the spacecraft can be synchronized at a common attitude with zero angular velocity.

Remark 11.9 *The sufficient condition in (11.48) implies that if an explicit function $\rho_i(\cdot)$ is found to satisfy (11.42), then the size of k_{in} can be determined accordingly. As shown in [142] and [102], $\rho_i(\cdot)$ is derived from χ_i in (11.41) according to the* *mean value theorem, which is a positive nondecreasing function and consists of the bounding functions for the partial derivatives of the elements in $H_i(\sigma_i)$ and $C_i(\sigma_i, \dot{\sigma}_i)$ with respect to σ_i and $\dot{\sigma}_i$. If the apriori knowledge of the system parameter J_i is available, the partial derivatives can be computed in theory. Then an explicit function $\rho_i(\cdot)$ can be found. However, since the inverse of $F_i(\sigma_i)$ is included in $H_i(\sigma_i)$ and $C_i(\sigma_i, \dot{\sigma}_i)$, it is not easy to derive the explicit expression of $\rho_i(\cdot)$. For convenience of notation, $\rho_i(\cdot)$ is expressed in an implicit function form. More details on $\rho_i(\cdot)$ can be found in [102]. Therefore, to satisfy the sufficient condition in (11.42), a sufficiently large k_{in} is required. Similar to [142] and [102], a suitable k_{in} can be chosen by trial and error.*

11.4 EXPERIMENT RESULTS

Consider a group of 4 spacecraft to illustrate the proposed distributed adaptive attitude synchronization schemes. The inertial matrices of these spacecraft are the same as those in [62], given in Table 11.1. Note that all the system parameters in the inertial matrices are unknown. The graph among the 4 spacecraft is shown in Figure 11.2.

TABLE 11.1
The Inertial Matrices of 4 Spacecraft

J_1	$[1, 0.1, 0.1; 0.1, 0.1, 0.1; 0.1, 0.1, 0.9]$ kg $\cdot$ m^2
J_2	$[1.5, 0.2, 0.3; 0.2, 0.9, 0.4; 0.3, 0.4, 2.0]$ kg $\cdot$ m^2
J_3	$[0.8, 0.1, 0.2; 0.1, 0.7, 0.3; 0.2, 0.3, 1.1]$ kg $\cdot$ m^2
J_4	$[1.2, 0.3, 0.7; 0.3, 0.9, 0.2; 0.7, 0.2, 1.4]$ kg $\cdot$ m^2

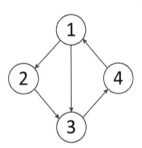

FIGURE 11.2 The communication graph among the 4 spacecraft.

In simulation, the initials including $\sigma_i(0)$, $\omega_i(0)$, $z_i(0)$, $z_{i,1}(0)$, $z_{i,2}(0)$, $z_{i,3}(0)$, and $\hat{\theta}_i(0)$ are set as $\sigma_1(0) = [0.5, 0.6, 0.7]^T$, $\sigma_2(0) = [0.4, 0.3, 0.2]^T$, $\sigma_3(0) =$

$[0.3, 0.4, 0.5]^T$, $\sigma_4(0) = [0.6, 0.7, 0.8]^T$, $\omega_i(0) = [0, 0, 0]^T$, $z_i(0) = z_{i,1}(0) = z_{i,2}(0) = z_{i,3}(0) = \sigma_i(0)$, $\hat{\theta}_i(0) = [0, 0, 0, 0, 0, 0]^T$ for $i = 1, \ldots, 4$.

11.4.1 ADAPTIVE ATTITUDE SYNCHRONIZATION WITH AVAILABLE ANGULAR VELOCITY MEASUREMENTS

The reference systems (11.5), triggering conditions (11.6), distributed adaptive control laws (11.21), (11.22) and parameter update laws (11.23) are adopted. The design parameters are chosen as $\pi_i = 0.5$, $\varsigma_i = 0.01$, $\iota_i = 0.6$, $K_i = 5I_3$, and $\Gamma_i = I_6$ for $i = 1, \ldots, 4$, where I_3 and I_6 are identity matrices with size 3×3 and size 6×6, respectively. The attitudes and angular velocities of the 4 spacecraft are provided in Figures 11.3–11.4, which shows that all the spacecraft can be synchronized at a common attitude with zero angular velocities. The control torques are shown in Figure 11.5. It can be observed from Figures 11.3–11.5 that all the observed signals are bounded. Moreover, Figure 11.6 shows that the Zeno behavior does not exist in each spacecraft.

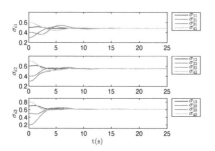

FIGURE 11.3 The attitude of spacecraft i for $i = 1, \ldots, 4$ in Case I.

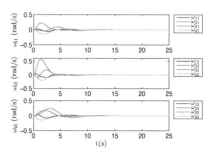

FIGURE 11.4 The angular velocity of spacecraft i for $i = 1, \ldots, 4$ in Case I.

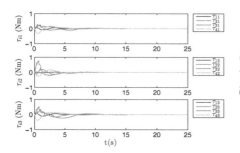

FIGURE 11.5 The control torque of spacecraft i for $i = 1, \ldots, 4$ in Case I.

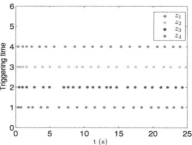

FIGURE 11.6 The triggering times of spacecraft i for $i = 1, \ldots, 4$ in Case I.

11.4.2 ADAPTIVE ATTITUDE SYNCHRONIZATION WITHOUT ANGULAR VELOCITY MEASUREMENTS

The reference systems (11.26)–(11.29), triggering conditions (11.30), pseudo-velocity filters (11.35)–(11.34), distributed adaptive output feedback control laws (11.43) and parameter update laws (11.45) are used. The triggering parameters π_i, ς_i, ι_i, and adaptation gain Γ_i are chosen the same as those in Case I. The design parameters k_{is}, $c_{i,1}$, and $c_{i,2}$ are chosen, respectively, as $k_{is} = 10$ and $c_{i,1} = c_{i,2} = 1$. The attitudes, angular velocities, control torques, and triggering times of the 4 spacecraft are respectively shown in Figures 11.7–11.10. It can be observed that the attitude synchronization of the 4 spacecraft is achieved and the angular velocities can converge to zero vector. Besides, the attitudes, angular velocities, and control torques are bounded and the Zeno behavior does not exist.

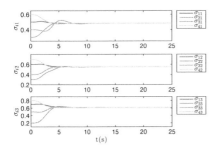

FIGURE 11.7 The attitude of spacecraft i for $i = 1, \ldots, 4$ in Case II.

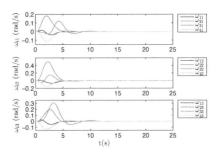

FIGURE 11.8 The angular velocity of spacecraft i for $i = 1, \ldots, 4$ in Case II.

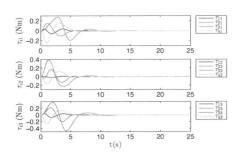

FIGURE 11.9 The control torque of spacecraft i for $i = 1, \ldots, 4$ in Case II.

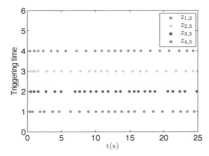

FIGURE 11.10 The triggering times of spacecraft i for $i = 1, \ldots, 4$ in Case II.

11.5 NOTES

In this chapter, the distributed adaptive attitude synchronization problem for multiple spacecraft with unknown inertial matrices and event-triggered communication has been considered under directed graph condition. Two distributed adaptive attitude synchronization schemes with and without angular velocity measurements are presented based on the designed reference systems. The triggering conditions in this chapter can not only reduce the communication burden but also avoid continuous monitoring of neighboring states, while the Zeno behavior in each spacecraft is rigorously ruled out. It is shown that all the closed-loop signals are uniformly bounded and the attitude of all the spacecraft can be synchronized at a common constant vector. Finally, the effectiveness of the proposed control schemes has been verified in simulation results.

References

1. A. Abdessameud and A. Tayebi. Attitude synchronization of a group of spacecraft without velocity measurements. *IEEE Transactions on Automatic Control*, 54(11):2642–2648, 2009.
2. M. Arcak. Passivity as a design tool for group coordination. *IEEE Transactions on Automatic Control*, 52(8):1380–1390, 2007.
3. K. J. Astrom and P. Eykhoff. System identification: A survey. *Automatica*, 7(2):123–162, 1971.
4. K. J. Astrom and B. Wittenmark. *Adaptive Control. 2nd ed.* Addison-Wesley, Reading, MA, 1995.
5. H. Bai, M. Arcak, and J. T. Wen. Adaptive design for reference velocity recovery in motion coordination. *Systems & Control Letters*, 57(8):602–610, 2008.
6. H. Bai, M. Arcak, and J. T. Wen. Adaptive motion coordination: Using relative velocity feedback to track a reference velocity. *Automatica*, 45(4):1020–1025, 2009.
7. T. Balch and R. C. Arkin. Behavior-based formation control for multirobot teams. *IEEE Transactions on Robotics and Automation*, 14(6):926–939, 1998.
8. R. W. Beard, J. Lawton, and F. Y. Hadaegh. A coordination architecture for spacecraft formation control. *IEEE Transactions on Control Systems Technology*, 9(6):777–790, 2001.
9. S. P. Bhat and D. Bernstein. Continuous finite-time stabilization of the translational and rotational double integrators. *IEEE Transactions on Automatic Control*, 43(5):678–682, 1998.
10. S. P. Bhat and D. Bernstein. Finite-time stability of continuous autonomous systems. *SIAM Journal on Control and Optimization*, 38(3):751–766, 2000.
11. Y. Cao, W. Yu, W. Ren, and G. Chen. An overview of recent progress in the study of distributed multi-agent coordination. *IEEE Transactions on Industrial Informatics*, 9(1):427–438, 2012.
12. C. P. Chen, G.X. Wen, Y.J. Liu, and Z. Liu. Observer-based adaptive backstepping consensus tracking control for high-order nonlinear semi-strict-feedback multiagent systems. *IEEE Transactions on Cybernetics*, 46(7):1591–1601, 2015.
13. N. Chopra and M. W. Spong. On exponential synchronization of kuramoto oscillators. *IEEE Transactions on Automatic Control*, 54(2):353–357, 2009.
14. S.J. Chung and J.J.E. Slotine. Cooperative robot control and concurrent synchronization of lagrangian systems. *IEEE Transactions on Robotics*, 25(3):686–700, 2009.
15. J. Cortés. Distributed algorithms for reaching consensus on general functions. *Automatica*, 44(3):726–737, 2008.
16. A. Das and F. Lewis. Distributed adaptive control for synchronization of unknown nonlinear networked systems. *Automatica*, 46(12):2014–2021, 2010.
17. A. Das and F. L. Lewis. Cooperative adaptive control for synchronization of second-order systems with unknown nonlinearities. *International Journal of Robust and Nonlinear Control*, 21(13):1509–1524, 2011-09.

18. D. V. Dimarogonas, E. Frazzoli, and K. Johansson. Distributed event-triggered control for multi-agent systems. *IEEE Transactions on Automatic Control*, 57(5):1291–1297, 2011.

19. D. V. Dimarogonas, E. Frazzoli, and K. H. Johansson. Distributed event-triggered control for multi-agent systems. *IEEE Transactions on Automatic Control*, 57(5):1291–1297, 2012.

20. D. V. Dimarogonas and K. J. Kyriakopoulos. On the state agreement problem for multiple unicycles. In *2006 American Control Conference*, pages 6 pp. 2016–2021, 2006.

21. D. V. Dimarogonas and K. J. Kyriakopoulos. On the rendezvous problem for multiple nonholonomic agents. *IEEE Transactions on Automatic Control*, 52(5):916–922, 2007.

22. D.V. Dimarogonas and K.J. Kyriakopoulos. On the state agreement problem for multiple unicycles with varying communication links. In *Proceedings of the 45th IEEE Conference on Decision and Control*, pages 4283–4288, 2006.

23. Z. Ding. Consensus disturbance rejection with disturbance observers. *IEEE Transactions on Industrial Electronics*, 62(9):5829–5837, 2015.

24. Z. Ding. Distributed adaptive consensus output regulation of network-connected heterogeneous unknown linear systems on directed graphs. *IEEE Transactions on Automatic Control*, 62(9):4683–4690, 2016.

25. K. D. Do and J. Pan. Nonlinear formation control of unicycle-type mobile robots. *Robotics and Autonomous Systems*, 55(3):191–204, 2007.

26. W. Dong. Adaptive consensus seeking of multiple nonlinear systems. *International Journal of Adaptive Control and Signal Processing*, 26(5):419–434, 2012.

27. W. Dong and J. A. Farrell. Consensus of multiple nonholonomic systems. In *2008 47th IEEE Conference on Decision and Control*, pages 2270–2275, 2008.

28. W. Dong and J. A. Farrell. Cooperative control of multiple nonholonomic mobile agents. *IEEE Transactions on Automatic Control*, 53(6):1434–1448, 2008.

29. W. Dong, M. B. Ghalia, and J. A. Farrell. Tracking control of multiple nonlinear systems via information interchange. In *2011 50th IEEE Conference on Decision and Control and European Control Conference*, pages 5076–5081, 2011.

30. S. El-Ferik, A. Qureshi, and F.L. Lewis. Neuro-adaptive cooperative tracking control of unknown higher-order affine nonlinear systems. *Automatica*, 50(3):798–808, 2014.

31. H. Elliott. Direct adaptive pole placement with application to nonminimum phase systems. *IEEE Transactions on Automatic Control*, 27(3):720–722, 1982.

32. H. Elliott, W. Wolovich, and M. Das. Arbitrary adaptive pole placement for linear multivariable systems. *IEEE Transactions on Automatic Control*, 29(3):221–229, 1984.

33. X. Ge, Q.L. Han, L. Ding, Y. Wang, and X. Zhang. Dynamic event-triggered distributed coordination control and its applications: A survey of trends and techniques. *IEEE Transactions on Systems, Man, and Cybernetics: Systems*, 50(9):3112–3125, 2020.

34. G. C. Goodwin, D. J. Hill, D. Q. Mayne, and R. H. Middleton. Adaptive robust control (convergence, stability and performance). In *Proceedings of the 25th IEEE CDC*, volume 1, pages 468–473, Athens, Greece, 1986.

35. G. C. Goodwin and K. S. Sin. *Adaptive Filtering Prediction and Control*. Prentice Hall, Englewook Cliffs, NJ, 1984.

36. X. Guo, G. Wei, M. Yao, and P. Zhang. Consensus control for multiple euler-lagrange systems based on high-order disturbance observer: An event-triggered approach. *IEEE/CAA Journal of Automatica Sinica*, 9(5):945–948, 2022.

37. Z. Han, Z. Lin, M. Fu, and Z. Chen. Distributed coordination in multi-agent systems: a graph laplacian perspective. *Frontiers of Information Technology & Electronic Engineering*, 16:429–448, 2015.

38. G. H. Hardy, J. E. Littlewood, and G. Pólya. *Inequalities*. Cambridge university press, 1952.

39. W. He and J. Cao. Consensus control for high-order multi-agent systems. *IET Control Theory & Applications*, 5(1):231–238, 2011.

40. Y. Hong, J. Hu, and L. Gao. Tracking control for multi-agent consensus with an active leader and variable topology. *Automatica*, 42(7):1177–1182, 2006.

41. Z.G. Hou, L. Cheng, and M. Tan. Decentralized robust adaptive control for the multi-agent system consensus problem using neural networks. *IEEE Transactions on Systems, Man, and Cybernetics, Part B (Cybernetics)*, 39(3):636–647, 2009.

42. G. Hu. Robust consensus tracking for an integrator-type multi-agent system with disturbances and unmodelled dynamics. *International Journal of Control*, 84(1):1–8, 2011.

43. J. Hu and W. Zheng. Adaptive tracking control of leader–follower systems with unknown dynamics and partial measurements. *Automatica*, 50(5):1416–1423, 2014.

44. W. Hu, L. Liu, and G. Feng. Output consensus of heterogeneous linear multi-agent systems by distributed event-triggered/self-triggered strategy. *IEEE Transactions on Cybernetics*, 47(8):1914–1924, 2016.

45. S. Huang, K. K. Tan, and T. H. Lee. Decentralized control design for large-scale systems with strong interconnections using neural networks. *IEEE Transactions on Automatic Control*, 48(5):805–810, 2003.

46. P. Ioannou and K. Tsakalis. A robust discrete-time adaptive controller. In *Proceedings of the 25th IEEE CDC*, volume 1, pages 838–843, Athens, Greece, 1986.

47. P. A. Ioannou and J. Sun. *Robust adaptive control*. PTR Prentice-Hall Upper Saddle River, NJ, 1996.

48. Petros A Ioannou and Jing Sun. *Robust adaptive control*, volume 1. PTR Prentice-Hall Upper Saddle River, NJ, 1996.

49. F. Jiang and L. Wang. Consensus seeking of high-order dynamic multi-agent systems with fixed and switching topologies. *International Journal of Control*, 83(2):404–420, 2010.

50. X. Jin. Adaptive iterative learning control for high-order nonlinear multi-agent systems consensus tracking. *Systems & Control Letters*, 89:16–23, 2016.

51. Y. Kaizuka and K. Tsumura. Consensus via distributed adaptive control. *IFAC Proceedings Volumes*, 44(1):1213–1218, 2011.

52. Y. Kim and M. Mesbahi. On maximizing the second smallest eigenvalue of a state-dependent graph laplacian. In *Proceedings of the 2005, American Control Conference, 2005*, pages 99–103, 2005.

53. G. Kreisselmeier and B. D. O. Anderson. Robust model reference adaptive control. *IEEE Transactions on Automatic Control*, AC-31(2):127–133, 1986.

54. E. Kreyszig, K. Stroud, and G. Stephenson. *Advanced engineering mathematics, 9th edition*. John Wiley & Sons, Inc., New York, 2008.

55. M. Krstić, I. Kanellakopoulos, and P. Kokotović. Adaptive nonlinear control without overparametrization. *Systems & Control Letters*, 19(3):177–185, 1992.

56. M. Krstić, I. Kanellakopoulos, and P. Kokotović. *Nonlinear and Adaptive Control Design*. Wiley, New York, 1995.

57. J. R. Lawton and R. Beard. Synchronized multiple spacecraft rotations. *Automatica*, 38(8):1359–1364, 2002.

58. F. L. Lewis, D. M. Dawson, and C. Abdallah. *Robot manipulator control: theory and practice*. CRC Press, 2003.

59. H. Li, X. Liao, T. Huang, and W. Zhu. Event-triggering sampling based leader-following consensus in second-order multi-agent systems. *IEEE Transactions on Automatic Control*, 60(7):1998–2003, 2014.

60. S. Li, H. Du, and X. Lin. Finite-time consensus algorithm for multi-agent systems with double-integrator dynamics. *Automatica*, 47(8):1706–1712, 2011.

61. Y.X. Li, G.H. Yang, and S. Tong. Fuzzy adaptive distributed event-triggered consensus control of uncertain nonlinear multiagent systems. *IEEE Transactions on Systems, Man, and Cybernetics: Systems*, 49(9):1777–1786, 2019.

62. Z. Li and Z. Duan. Distributed adaptive attitude synchronization of multiple spacecraft. *Science China Technological Sciences*, 54:1992–1998, 2011.

63. Z. Li, Z. Duan, G. Chen, and L. Huang. Consensus of multiagent systems and synchronization of complex networks: A unified viewpoint. *IEEE Transactions on Circuits and Systems I: Regular Papers*, 57(1):213–224, 2009.

64. Z. Li, X. Liu, W. Ren, and L. Xie. Distributed tracking control for linear multiagent systems with a leader of bounded unknown input. *IEEE Transactions on Automatic Control*, 58(2):518–523, 2012.

65. Z. Li, W. Ren, X. Liu, and M. Fu. Consensus of multi-agent systems with general linear and Lipschitz nonlinear dynamics using distributed adaptive protocols. *IEEE Transactions on Automatic Control*, 58(7):1786–1791, 2012.

66. Z. Li, G. Wen, Z. Duan, and W. Ren. Designing fully distributed consensus protocols for linear multi-agent systems with directed graphs. *IEEE Transactions on Automatic Control*, 60(4):1152–1157, 2014.

67. Z. Li, J. Yan, W. Yu, and J. Qiu. Adaptive event-triggered control for unknown second-order nonlinear multiagent systems. *IEEE Transactions on Cybernetics*, 51(12):6131–6140, 2021.

68. P. Lin and Y. Jia. Robust H_∞ consensus analysis of a class of second-order multi-agent systems with uncertainty. *IET Control Theory & Applications*, 3(4):487–498, 2010.

69. P. Lin, Y. Jia, and L. Li. Distributed robust H_∞ consensus control in directed networks of agents with time-delay. *Systems & Control Letters*, 57(8):643–653, 2008.

70. Z. Lin, B. Francis, and M. Maggiore. Necessary and sufficient graphical conditions for formation control of unicycles. *IEEE Transactions on Automatic Control*, 50(1):121–127, 2005.

71. J. Liu, Y. Zhang, Y. Yu, and C. Sun. Fixed-time event-triggered consensus for nonlinear multiagent systems without continuous communications. *IEEE Transactions on Systems, Man, and Cybernetics: Systems*, 49(11):2221–2229, 2018.

72. Y. Liu and Y. Jia. Robust H_∞ consensus control of uncertain multi-agent systems with time delays. *International Journal of Control, Automation and Systems*, 9(6):1086–1094, 2011.

73. Y. Liu and Y. Jia. Adaptive consensus protocol for networks of multiple agents with nonlinear dynamics using neural networks. *Asian Journal of Control*, 14(5):1328–1339, 2012.

74. J. Long, W. Wang, J. Huang, J. Zhou, and K. Liu. Distributed adaptive control for asymptotically consensus tracking of uncertain nonlinear systems with intermittent actuator faults and directed communication topology. *IEEE Transactions on Cybernetics*, 51(8):4050–4061, 2021.

75. J. Lü, F. Chen, and G. Chen. Nonsmooth leader-following formation control of nonidentical multi-agent systems with directed communication topologies. *Automatica*, 64:112–120, 2016.

76. L. Ma, H. Min, S. Wang, Y. Liu, and S. Liao. An overview of research in distributed attitude coordination control. *IEEE/CAA Journal of Automatica Sinica*, 2(2):121–133, 2015.

77. J. Mei, W. Ren, and G. Ma. Distributed coordinated tracking with a dynamic leader for multiple euler-lagrange systems. *IEEE Transactions on Automatic Control*, 56(6):1415–1421, 2011.

78. J. Mei, W. Ren, and G. Ma. Distributed containment control for lagrangian networks with parametric uncertainties under a directed graph. *Automatica*, 48(4):653–659, 2012.

79. Z. Meng, W. Ren, and Z. You. Distributed finite-time attitude containment control for multiple rigid bodies. *Automatica*, 46(12):2092–2099, 2010.

80. G. Miao, S. Xu, and Y. Zou. Consentability for high-order multi-agent systems under noise environment and time delays. *Journal of the Franklin Institute*, 350(2):244–257, 2013.

81. H. Min, S. Wang, F. Sun, Z. Gao, and J. Zhang. Decentralized adaptive attitude synchronization of spacecraft formation. *Systems & Control Letters*, 61(1):238–246, 2012.

82. R. Monopoli. Model reference adaptive control with an augmented error signal. *IEEE Transactions on Automatic Control*, 19(5):474–484, 1974.

83. L. Moreau. Stability of multiagent systems with time-dependent communication links. *IEEE Transactions on Automatic Control*, 50(2):169–182, 2005.

84. P. Morin and C. Samson. Practical stabilization of driftless systems on lie groups: the transverse function approach. *IEEE Transactions on Automatic Control*, 48(9):1496–1508, 2003.

85. S. M. Naik, P. R. Kumar, and B. E. Ydstie. Robust continuous-time adaptive control by parameter projection. *IEEE Transactions on Automatic Control*, 37(2):182–197, 1992.

86. K. S. Narendra and A. M. Annaswamy. *Stable Adaptive Systems*. Prentice Hall, Englewook Cliffs, NJ, 1989.

87. W. Ni and D. Cheng. Leader-following consensus of multi-agent systems under fixed and switching topologies. *Systems & Control Letters*, 59(3-4):209–217, 2010.

88. C. Nowzari and J. Cortés. Distributed event-triggered coordination for average consensus on weight-balanced digraphs. *Automatica*, 68:237–244, 2016.

89. C. Nowzari, E. Garcia, and J. Cortés. Event-triggered communication and control of networked systems for multi-agent consensus. *Automatica*, 105:1–27, 2019.

90. E. Nuno, R. Ortega, L. Basanez, and D. Hill. Synchronization of networks of nonidentical euler-lagrange systems with uncertain parameters and communication delays. *IEEE Transactions on Automatic Control*, 56(4):935–941, 2011.

91. R. Olfati-Saber and R. Murray. Consensus problems in networks of agents with switching topology and time-delays. *IEEE Transactions on Automatic Control*, 49(9):1520–1533, 2004.

92. R. Ortega, L. Praly, and I. D. Landau. Robustness of discrete-time direct adaptive controllers. *IEEE Transactions on Automatic Control*, AC-30(12):1179–1187, 1985.

93. P. C. Parks. Liapunov redesign of model reference adaptive control systems. *IEEE Transactions on Automatic Control*, 11(3):362–367, 1966.

94. I. S. Parry and C. H. Houpis. A parameter identification self-adaptive control system. *IEEE Transactions on Automatic Control*, 15(4):426–428, 1970.

95. J.B. Pomet and L. Praly. Adaptive nonlinear regulation: Estimation from the lyapunov equation. *IEEE Transactions on Automatic Control*, 37(6):729–740, 1992.

96. L. Praly. Robustness of model reference adaptive control. In *Proceedings of the 3rd Yale Workshop on Applications of Adaptive Systems Theory*, pages 224–226, 1983.

97. C. Qian and W. Lin. A continuous feedback approach to global strong stabilization of nonlinear systems. *IEEE Transactions on Automatic Control*, 46(7):1061–1079, 2001.

98. C. Qian and W. Lin. Non-Lipschitz continuous stabilizers for nonlinear systems with uncontrollable unstable linearization. *Systems & Control Letters*, 42(3):185–200, 2001.

99. Y. Qian, L. Liu, and G. Feng. Distributed event-triggered adaptive control for consensus of linear multi-agent systems with external disturbances. *IEEE Transactions on Cybernetics*, 50(5):2197–2208, 2018.

100. J. Qin, Q. Ma, Y. Shi, and L. Wang. Recent advances in consensus of multi-agent systems: A brief survey. *IEEE Transactions on Industrial Electronics*, 64(6):4972–4983, 2017-06.

101. J. Qin, Q. Ma, W. X. Zheng, H. Gao, and Y. Kang. Robust H_∞ group consensus for interacting clusters of integrator agents. *IEEE Transactions on Automatic Control*, 62(7):3559–3566, 2017.

102. M. S. Queiroz, J. Hu, D. M. Dawson, T. Burg, and S. Donepudi. Adaptive position/force control of robot manipulators without velocity measurements: Theory and experimentation. *IEEE Transactions on Systems, Man, and Cybernetics, Part B (Cybernetics)*, 27(5):796–809, 1997.

103. C.-E. Ren and C. P. Chen. Sliding mode leader-following consensus controllers for second-order non-linear multi-agent systems. *IET Control Theory & Applications*, 9(10):1544–1552, 2015.

104. W. Ren. Consensus strategies for cooperative control of vehicle formations. *IET Control Theory & Applications*, 1(2):505–512, 2007.

105. W. Ren. Multi-vehicle consensus with a time-varying reference state. *Systems & Control Letters*, 56(7-8):474–483, 2007.

106. W. Ren. Distributed cooperative attitude synchronization and tracking for multiple rigid bodies. *IEEE Transactions on Control Systems Technology*, 18(2):383–392, 2009.

107. W. Ren. Distributed leaderless consensus algorithms for networked euler–lagrange systems. *International Journal of Control*, 82(11):2137–2149, 2009.

108. W. Ren and E. Atkins. Distributed multi-vehicle coordinated control via local information exchange. *International Journal of Robust and Nonlinear Control: IFAC-Affiliated Journal*, 17(10-11):1002–1033, 2007.

109. W. Ren and R. Beard. Consensus seeking in multiagent systems under dynamically changing interaction topologies. *IEEE Transactions on Automatic Control*, 50(5):655–661, 2005.

110. W. Ren, R. W. Beard, and E. M. Atkins. A survey of consensus problems in multi-agent coordination. In *Proceedings of the 2005, American Control Conference, 2005*, pages 1859–1864, 2005.

111. W. Ren and Y. Cao. *Distributed Coordination of Multi-Agent Networks: Emergent Problems, Models and Issues*. Springer, London, 2010.

112. W. Ren and Y. Cao. *Distributed coordination of multi-agent networks: emergent problems, models, and issues*, volume 1. Springer, 2011.

113. W. Ren, K. Moore, and Y. Chen. High-order consensus algorithms in cooperative vehicle systems. In *2006 IEEE International Conference on Networking, Sensing and Control*, pages 457–462, 2006.

114. W. Ren, K. L. Moore, and Y. Chen. High-order and model reference consensus algorithms in cooperative control of multi-vehicle systems. *Journal of Dynamic Systems, Measurement, and Control*, 129(5):678–688, 2006.

115. A. Sarlette, R. Sepulchre, and N. E. Leonard. Autonomous rigid body attitude synchronization. *Automatica*, 45(2):572–577, 2009.

116. J. H. Seo, H. Shim, and J. Back. Consensus of high-order linear systems using dynamic output feedback compensator: Low gain approach. *Automatica*, 45(11):2659–2664, 2009.

117. G. S. Seyboth, D. V. Dimarogonas, and K. Johansson. Event-based broadcasting for multi-agent average consensus. *Automatica*, 49(1):245–252, 2013.

118. G. S. Seyboth, D. V. Dimarogonas, and K. H. Johansson. Event-based broadcasting for multi-agent average consensus. *Automatica*, 49(1):245–252, 2013.

119. Q. Shen and P. Shi. Distributed command filtered backstepping consensus tracking control of nonlinear multiple-agent systems in strict-feedback form. *Automatica*, 53:120–124, 2015.

120. M. W. Spong and M. Vidyasagar. *Robot dynamics and control*. John Wiley & Sons, 2008.

121. H. Su, G. Chen, X. Wang, and Z. Lin. Adaptive second-order consensus of networked mobile agents with nonlinear dynamics. *Automatica*, 47(2):368–375, 2011.

122. H. Su, X. Wang, and Z. Lin. Flocking of multi-agents with a virtual leader. *IEEE Transactions on Automatic Control*, 54(2):293–307, 2009.

123. Z. Sun, N. Huang, B. D. Anderson, and Z. Duan. Event-based multiagent consensus control: Zeno-free triggering via $\mathcal{L}^p$ signals. *IEEE Transactions on Cybernetics*, 50(1):284–296, 2018.

124. P. Tabuada. Event-triggered real-time scheduling of stabilizing control tasks. *IEEE Transactions on Automatic Control*, 52(9):1680–1685, 2007.

125. G. Tao. *Adaptive Control Design and Analysis*. John Wiley & Sons, Inc., New York, 2003.

126. S. E. Tuna. Lqr-based coupling gain for synchronization of linear systems. *Mathematics*, 2008.

127. S. E. Tuna. Conditions for synchronizability in arrays of coupled linear systems. *IEEE Transactions on Automatic Control*, 54(10):2416–2420, 2009.

128. J. Wang. Distributed coordinated tracking control for a class of uncertain multiagent systems. *IEEE Transactions on Automatic Control*, 62(7):3423–3429, 2016.

129. L. Wang and F. Xiao. Finite-time consensus problems for networks of dynamic agents. *IEEE Transactions on Automatic Control*, 55(4):950–955, 2010.

130. W. Wang, J. Huang, C. Wen, and H. Fan. Distributed adaptive control for consensus tracking with application to formation control of nonholonomic mobile robots. *Automatica*, 50(4):1254–1263, 2014.

131. W. Wang and Y. Li. Observer-based event-triggered adaptive fuzzy control for leader-following consensus of nonlinear strict-feedback systems. *IEEE Transactions on Cybernetics*, 51(4):2131–2141, 2021.

132. W. Wang, J. Long, C. Wen, and J. Huang. Recent advances in distributed adaptive consensus control of uncertain nonlinear multi-agent systems. *Journal of Control and Decision*, 7(1):44–63, 2020.

133. W. Wang, C. Wen, and J. Huang. Distributed adaptive asymptotically consensus tracking control of nonlinear multi-agent systems with unknown parameters and uncertain disturbances. *Automatica*, 77:133–142, 2017.

134. W. Wang, C. Wen, J. Huang, and H. Fan. Distributed adaptive asymptotically consensus tracking control of uncertain euler-lagrange systems under directed graph condition. *ISA Transactions*, 71:121–129, 2017.

135. W. Wang, C. Wen, J. Huang, and H. Fan. Distributed adaptive asymptotically consensus tracking control of uncertain euler-lagrange systems under directed graph condition. *ISA Transactions*, 71:121–129, 2017.

136. W. Wang, C. Wen, J. Huang, and Z. Li. Hierarchical decomposition based consensus tracking for uncertain interconnected systems via distributed adaptive output feedback control. *IEEE Transactions on Automatic Control*, 61(7):1938–1945, 2015.

137. W. Wang, C. Wen, J. Huang, and J. Zhou. Adaptive consensus of uncertain nonlinear systems with event triggered communication and intermittent actuator faults. *Automatica*, 111:108667, 2020.

138. X. Wang, V. Yadav, and S. N. Balakrishnan. Cooperative uav formation flying with obstacle/collision avoidance. *IEEE Transactions on Control Systems Technology*, 15(4):672–679, 2007.

139. Z. Wang, W. Zhang, and Y. Guo. Adaptive output consensus tracking of uncertain multi-agent systems. In *Proceedings of the 2011 American Control Conference*, pages 3387–3392, 2011.

140. C. Wen and D. J. Hill. Robustness of adaptive control without deadzones, data normalization or persistence of excitation. *Automatica*, 25(6):943–947, 1989.

141. H. P. Whitaker, J. Yamron, and A. Kezer. Design of model reference adaptive control systems for aircraft. *Report R-164, Instrumentation Laboratory, Massachusetts Institute of Technology*, 1958.

142. H. Wong, M. S. Queiroz, and V. Kapila. Adaptive tracking control using synthesized velocity from attitude measurements. *Automatica*, 37(6):947–953, 2001.

143. F. Xiao, L. Wang, J. Chen, and Y. Gao. Finite-time formation control for multi-agent systems. *Automatica*, 45(11):2605–2611, 2009.

144. L. Xiao and S. Boyd. Fast linear iterations for distributed averaging. *Systems & Control Letters*, 53(1):65–78, 2004.

145. D. Xie, S. Xu, Y. Chu, and Y. Zou. Event-triggered average consensus for multi-agent systems with nonlinear dynamics and switching topology. *Journal of the Franklin Institute*, 352(3):1080–1098, 2015.

146. L. Xing, C. Wen, F. Guo, Z. Liu, and H. Su. Event-based consensus for linear multiagent systems without continuous communication. *IEEE Transactions on Cybernetics*, 47(8):2132–2142, 2016.

147. L. Xing, C. Wen, F. Guo, Z. Liu, and H. Su. Event-based consensus for linear multiagent systems without continuous communication. *IEEE Transactions on Cybernetics*, 47(8):2132–2142, 2017.

148. H. Yang, Z. Zhang, and S. Zhang. Consensus of second-order multi-agent systems with exogenous disturbances. *International Journal of Robust and Nonlinear Control*, 21(9):945–956, 2011.

149. T. Yang, Y. Jin, W. Wang, and Y. Shi. Consensus of high-order continuous-time multi-agent systems with time-delays and switching topologies. *Chinese Physics B*, 20(2):020511, 2011.

150. Y. Yang, Y. Li, D. Yue, and W. Yue. Adaptive event-triggered consensus control of a class of second-order nonlinear multiagent systems. *IEEE Transactions on Cybernetics*, 50(12):5010–5020, 2019.

151. Y. Yang, Y. Li, D. Yue, and W. Yue. Adaptive event-triggered consensus control of a class of second-order nonlinear multiagent systems. *IEEE Transactions on Cybernetics*, 50(12):5010–5020, 2020.

152. B. E. Ydstie. Stability of discrete model reference adaptive control – revisited. *System & Control Letters*, 13(5):429–438, 1989.

153. S. J. Yoo. Distributed consensus tracking for multiple uncertain nonlinear strict-feedback systems under a directed graph. *IEEE Transactions on Neural Networks and Learning Systems*, 24(4):666–672, 2013.

154. S. J. Yoo. Connectivity-preserving consensus tracking of uncertain nonlinear strict-feedback multiagent systems: An error transformation approach. *IEEE Transactions on Neural Networks and Learning Systems*, 29(9):4542–4548, 2017.

155. H. Yu and X. Xia. Adaptive consensus of multi-agents in networks with jointly connected topologies. *Automatica*, 48(8):1783–1790, 2012.

156. W. Yu, G. Chen, and M. Cao. Some necessary and sufficient conditions for second-order consensus in multi-agent dynamical systems. *Automatica*, 46(6):1089–1095, 2010.

157. W. Yu, G. Chen, W. Ren, J. Kurths, and W. Zheng. Distributed higher order consensus protocols in multiagent dynamical systems. *IEEE Transactions on Circuits and Systems I: Regular Papers*, 58(8):1924–1932, 2011.

158. W. Yu, H. Wang, F. Cheng, X. Yu, and G. Wen. Second-order consensus in multiagent systems via distributed sliding mode control. *IEEE Transactions on Cybernetics*, 47(8):1872–1881, 2016.

159. J. Zhan, Y. Hu, and X. Li. Adaptive event-triggered distributed model predictive control for multi-agent systems. *Systems & Control Letters*, 134:104531, 2019.

160. H. Zhang, G. Feng, H. Yan, and Q. Chen. Observer-based output feedback event-triggered control for consensus of multi-agent systems. *IEEE Transactions on Industrial Electronics*, 61(9):4885–4894, 2013.

161. H. Zhang and F. L. Lewis. Adaptive cooperative tracking control of higher-order nonlinear systems with unknown dynamics. *Automatica*, 48(7):1432–1439, 2012.

162. H. Zhang, F. L. Lewis, and Z. Qu. Lyapunov, adaptive, and optimal design techniques for cooperative systems on directed communication graphs. *IEEE Transactions on Industrial Electronics*, 59(7):3026–3041, 2011.

163. H. Zhang, Z. Li, Z. Qu, and F. Lewis. On constructing Lyapunov functions for multi-agent systems. *Automatica*, 58:39–42, 2015.

164. J. Zhang, K.H. Johansson, J. Lygeros, and S. Sastry. Zeno hybrid systems. *International Journal of Robust and Nonlinear Control*, 11(5):435–451, 2001.

165. W. Zhang, D. Zeng, and S. Qu. Dynamic feedback consensus control of a class of high-order multi-agent systems. *IET Control Theory & Applications*, 10(4):2219–2222, 2010.

166. Y. Zhang and Y. Yang. Finite-time consensus of second-order leader-following multi-agent systems without velocity measurements. *Physics letters A*, 377(3-4):243–249, 2013.

167. D. Zhao, T. Zou, S. Li, and Q. Zhu. Adaptive backstepping sliding mode control for leader-follower multi-agent systems. *IET Control Theory & Applications*, 6(8):1109–1117, 2012.

168. J. Zhou and C. Wen. *Adaptive backstepping control of uncertain systems: Nonsmooth nonlinearities, interactions or time-variations*. Springer, 2008.

169. J. Zhou, C. Wen, W. Wang, and F. Yang. Adaptive backstepping control of nonlinear uncertain systems with quantized states. *IEEE Transactions on Automatic Control*, 64(11):4756–4763, 2019.

170. W. Zhu, Z.P. Jiang, and G. Feng. Event-based consensus of multi-agent systems with general linear models. *Automatica*, 50(2):552–558, 2014.

171. W. Zhu, Z.P. Jiang, and G. Feng. Event-based consensus of multi-agent systems with general linear models. *Automatica*, 50(2):552–558, 2014.

172. Z. Zuo and C. Wang. Adaptive trajectory tracking control of output constrained multi-rotors systems. *IET Control Theory & Applications*, 8(13):1163–1174, 2014.

Index

For Product Safety Concerns and Information please contact our EU
representative GPSR@taylorandfrancis.com
Taylor & Francis Verlag GmbH, Kaufingerstraße 24, 80331 München, Germany